APPROACHES TO SEMIOTICS

edited by

THOMAS A. SEBEOK

Research Center for the Language Sciences, Indiana University

27

THE EXISTENTIAL GRAPHS OF CHARLES S. PEIRCE

by

DON D. ROBERTS

University of Waterloo

1973

MOUTON

THE HAGUE · PARIS

LIBRARY OF CONGRESS CATALOG CARD NUMBER: 73-85776

Printed in The Netherlands

For
Max Fisch
Ben dell' Intelletto

BEN DELL' INTELLETTO

Whenever *Good of Intellect* comes in,
Then peace is with us, and a soft control
Of all harsh thinking; and but one desire
Fills every bosom, –to forget the din
Of outside things, and render up the soul
To friendship's banquet by an evening fire.
Then is the season in this world of sin
That brings new strength, and keepeth us heart-whole
Amid the changes that distress and tire;
And when from wisdom we have wanderers been,
So that a stupor on the spirit stole
*From things unknown,** with visions dark and dire,
In this high presence we restore ourselves
More than by all the volumes on our shelves.

T.W. Parsons

* "E stupor m' eran le cose non conte." –*Purgatorio*, xv. 12.

PREFACE

The primary sources for this study are the eight volume *Collected Papers of Charles Sanders Peirce* and the unpublished manuscripts in the Houghton Library at Harvard University. Reference to the *Collected Papers* (*CP*) will be by the now standard decimal notation according to which, for example, the number 4.372 refers to paragraph 372 of volume 4 of *CP*. Reference to the unpublished manuscripts, with one exception, will be by the numbers given in Richard Robin's *Annotated Catalogue* and 'Supplementary Catalogue'; these numbers will occur with the prefix 'Ms'. An extra 's' in the number of an unpublished manuscript–as in 'Ms 499(s)' and 'Ms S 27'–will indicate that the item is listed in Robin's 'Supplementary Catalogue'. The one exception to this method is Ms 339, a notebook whose dated entries range from 1865 to 1909; since it is frequently useful to identify this manuscript as the source of a quotation or reference, I will ordinarily refer to it by the letters '*LN*' for '*Logic Notebook*'.

Most of the material on the graphs is to be found in the unpublished papers. Here there are more than thirty different expositions of systems of logical graphs, and here there is frequent use of graphs to introduce or clarify some logical or mathematical or philosophical concept. The exposition given below is the result of an attempt to distill out of this plenitude–plus the material in *CP*–a single adequate presentation of the system of existential graphs which would include the major and lasting revisions which Peirce introduced from time to time. I do not claim to have exhausted the topic, nor has that been my intention. My aim has been, rather, to make this material more accessible to students of Peirce who, for one reason or another, have not yet directed their attention to this important part of his work.

Max Fisch has been in on this project since 1960, when it began as a dissertation under his direction. He made it possible for me to work with the Peirce papers for extended periods of time; he has been unfailingly generous with his time and with results of his own research whenever I called for help; and many of his penetrating observations were accompanied by suggestions that are now incorporated into the text. But it is not for these reasons, sufficient though they are, that I dedicate the book to him. The reasons for that are expressed in the poem by Parsons. It was written for Benjamin Peirce, Charles's father. I should like to have written it myself for Max Fisch.

Several members of the University of Waterloo, friends and students of mine, have assisted me with the final revisions of this book. Beverley Kent in particular contributed to the major overhaul of the entire manuscript. E. James Crombie and Christopher P. Gray were of help especially in the more strictly logical matters, both causing important improvements in several places, such as in the statement of rules of transformation. Lorraine C. Beattie, who proof-read all of it, and Charles J. Dymer made many valuable suggestions and caught some ambiguities that had actively resisted discovery. Chapters 5 and 6, which cost me months of concentrated effort, owe much to the work of Mr. Gray and Miss Beattie. No doubt errors and ambiguities remain.

Throughout the book, special symbols and even letters of the alphabet drafted for technical use will be used as names of themselves wherever no ambiguity results from such usage. To assist the reader with terminology, a glossary is given in Appendix 5. Full information on the books and articles mentioned and quoted can be found in the bibliography. The dedicatory poem is in the Houghton Library, reference number *fAC85.P2566.B891. On a sheet dated December 23, 1872, Parsons wrote:" 'Ben dell' Intelletto' in Dante's Inferno meant Intellectual Good; and Benjamin Peirce, CSP's father, we called 'Ben dell' Intelletto'."

I wish to express my appreciation to the Department of Philosophy at Harvard University for permission to examine the Peirce papers in the Houghton Library and to quote extensively from them. Quotations from *Collected Papers of Charles Sanders Peirce*, edited by Charles Hartshorne, Paul Weiss, and Arthur W. Surks, Vols. I-VIII, are made with the kind permission of The Belknap Press of Harvard University Press, and are copyright 1931, '32, '33, '34, '35, '58, '59, '60, '61, '62, and '63 by the President and Fellows of Harvard College.

This book has been published with the help of a grant from the Humanities Research Council of Canada, using funds provided by the Canada Council.

TABLE OF CONTENTS

PREFACE 7

1 INTRODUCTION 11

2 THE DEVELOPMENT OF EXISTENTIAL GRAPHS 16
- 2.1 Peirce's Early Emphasis on Diagrammatic Thinking and Analysis 16
- 2.2 The Earliest Applications of the Diagrammatic Method to the Logic of Relatives 17
- 2.3 The Influence of Kempe 20
- 2.4 The First System: The Entitative Graphs 25
- 2.5 The Second System: The Existential Graphs 27

3 ALPHA 31
- 3.1 The Alpha Conventions 31
- 3.2 The Alpha Rules of Transformation 40
- 3.3 Further Illustrations 45

4 BETA 47
- 4.1 The Beta Conventions 47
- 4.2 The Beta Rules of Transformation 56
- 4.3 Further Illustrations 60

5 GAMMA 64
- 5.1 The 1903 Account of Abstractions 64
  - 5.11 The Potentials 68
  - 5.12 Graphs of Graphs 71
  - 5.13 The Three Gamma Rims 75
- 5.2 The 1903 Account of Modality 80

6 TINCTURED EXISTENTIAL GRAPHS 87
6.1 The Tinctures 88
6.2 Prolegomena 92
6.21 On Behalf of the Tinctures 95
6.22 1906 and After 98
6.3 Second Generation Tinctures 104
6.31 Conventions 105
6.32 Rules 107

7 GRAPHICAL ANALYSIS AND OUTCROPPINGS 110
7.1 The Purpose of Existential Graphs 110
7.11 The Graphical Analysis of Inference 111
7.12 The Graphical Analysis of Propositions 113
7.121 Subject and Predicate 114
7.122 Irreducibility of Triads 115
7.123 Composition of Concepts 116
7.2 Outcroppings 118
7.21 The Graphical Notation 118
7.22 EG Axioms and Natural Deduction 119
7.23 The Categorical Syllogism 122
7.24 Iconicity 123
7.25 Postscript 126

APPENDIX 1. A Selective Chronology of Peirce's Work on Logic 129

APPENDIX 2. Table of Logic Notations 136

APPENDIX 3. EG Conventions and Rules 137

APPENDIX 4. Completeness and Consistency 139
1. Alpha 139
2. Beta 145

APPENDIX 5. Glossary 152

BIBLIOGRAPHY 155

SUBJECT INDEX 159

NAME INDEX 163

REFERENCES TO THE COLLECTED PAPERS OF PEIRCE 164

REFERENCES TO THE PEIRCE MANUSCRIPTS 167

1

INTRODUCTION

Above the other titles he might justly have claimed, Charles S. Peirce prized the title 'logician'. He expressed in several places his confidence that his own special talent lay in the direction of logical analysis.[1] Indeed, in the judgment of some of the best authorities in the field today, he was a logician of great originality and power.[2]

Late in 1896[3] Peirce invented a system of logic diagrams which he soon began to call the 'existential graphs' ('EG' abbreviates 'the system of existential graphs'). He was 57 years old at the time, and had already made what scholars take to be his most significant contributions to modern symbolic logic.[4] In fact, his interest in the graphs grew, to a large extent, out of his pioneering work on symbolic logic. As he developed the system in the years after 1896 he began to find it relevant to such topics as modality, the theory of signs, his doctrine of the categories, and his philosophy of pragmaticism. By 1905 Peirce claimed that EG was "quite the luckiest find that has been gained in exact logic since Boole";[5] in 1908 he called it—or the theory of logical analysis which he based on it—his *'chef d'oeuvre'*;[6] and in 1909 he wrote to William James that EG "ought to be the logic of the future".[7]

In view of these statements by Peirce, it is interesting that no one has agreed with him regarding the value of his graphs. Perhaps this should not be

1 See for example Ms 631, 'Meaning, Preface', pp. 2ff, dated August 20, 1909. Cf. Peirce (1953), 37.
2 Sufficient evidence for this can be found in Appendix 1.
3 Ms 498; Ms 500; a letter from Peirce to F. A. Woods, begun October 4, 1913, in Ms L 477; cf. Ms 1589.
4 See note 2.
5 Ms 280, p. 22, a 1905-1906 manuscript on pragmaticism.
6 The phrase occurs in a letter to Philip E. B. Jourdain (so identified by Max H. Fisch), dated December 5, 1908. Hence the letter is a possible source for the quotation on p. 291 of *CP* 4. The passage is ambiguous, however, and needs further comment, which is given at the end of Chapter 7 below.
7 The date is December 25, 1909. Ms L 224.

surprising. Peirce's contemporaries had the advantage of some popular lectures on the graphs (the Lowell Lectures of 1903, principally), but his graphical publications were few and not easy to understand, as he admitted himself. What was needed was practice with the system, the kind of drill one gets in a classroom. And Peirce was not given the opportunity to provide this. When the volumes of the *Collected Papers* were being prepared, the editors had to choose from a large number of manuscripts and partial manuscripts which revealed several stages of development in EG. Quite properly they selected what Peirce had published, and a good bit of important material besides, but some of the simpler and more informal accounts were not included. The result has been that many accounts of Peirce's work simply ignore EG. James K. Feibleman, in his *An Introduction to Peirce's Philosophy*, writes at some length regarding Peirce's analysis of the nature of mathematical reasoning. In this connection he points out the importance Peirce attached to diagrammatic thinking;[8] and although this would seem to be a natural place to mention the logic diagrams, no explicit reference to EG occurs. There is explicit reference to EG in Murray G. Murphey's *The Development of Peirce's Philosophy*, and there are other passages which suggest the system.[9] Yet Murphey gives no indication that the graphs played any part in, or were at all related to, Peirce's philosophical development.

Even in works which concentrate on Peirce's contributions to formal logic, little is said of his work on logic diagrams. George D. W. Berry makes no mention of it at all in his article "Peirce's Contributions to the Logic of Statements and Quantifiers". Alonzo Church's *Introduction to Mathematical Logic* contains a wealth of historical material in special sections and in footnotes. He makes numerous references to Peirce, but none to EG. Two recent volumes which must be consulted in any treatment of the history of logic are I. M. Bochenski's *Formal Logic* and William and Martha Kneale's *The Development of Logic*. Bochenski's book is essentially a source book, the Kneale volume is essentially exposition; they complement each other admirably. Both books treat of Peirce's work in several connections, but there is no mention of his work on logic diagrams.

Authors who do acknowledge the existence of EG consider it too complicated to be of any value. W. V. O. Quine was one of the first to mention the graphs, and he did so in his 1934 review of *CP* 4. There he notes an important theme emphasized by Peirce, that the graphs were intended to

[8] Feibleman (1946), 108-109, 135-143. Let me state at the beginning, and once and for all, that these remarks must not be interpreted as wholesale criticism of the authors cited. On the contrary, I benefit greatly from their works, which I value highly. My task is far easier than theirs was, since I take for my topic a part only of Peirce's huge production.

[9] Murphey (1961), 399. See also 197n.22, 310n.12, and 342.

facilitate the analysis of logical structure, but not necessarily to facilitate the drawing of inferences. Peirce's insistence that his graphs were not designed to function as a calculus will be discussed in a later chapter. But Quine, after claiming that EG would be too cumbersome to function as a calculus, is not even optimistic about the analytical value of the graphs:

One questions the efficacy of Peirce's diagrams, however, in their analytical capacity as well. Their basic machinery is too complex to allow one much satisfaction in analyzing propositional structure into terms of that machinery. While it is not inconceivable that advances in the diagrammatic method might open possibilities of analysis superior to those afforded by the algebraic method, yet an examination of Peirce's product tends rather, apagogically as it were, to confirm one's faith in the algebraic approach [Quine (1934), 552].

Quine's opinions are everywhere worth serious consideration, and the fate of diagrams generally in the progress of logic since Peirce seems initially to confirm this judgment. Aside from the diagrams used in elementary logic texts to illustrate the categorical syllogism, logic graphs are rarely used today.

The case against the graphs can be argued from a somewhat different point of view. Berry, for example, in the article cited above, claims that Peirce made little or no contribution to either the logic of statements or the logic of quantifiers after 1891 (Berry [1952], 158-59, 165). Now if Berry is right, then, since Peirce discovered EG in 1896, EG contributes little or nothing to these areas of logic. Again, W. B. Gallie, in his *Peirce and Pragmatism*, claims that after 1891 Peirce's writing "like his thought, becomes looser: sometimes it rambles, at other times it becomes excessively disjointed" (Gallie [1952], 55). The tendencies, he says, are even plainer in certain papers written after 1900, papers in which

sense of shape, sense of proportion, and sustained unity of direction are often entirely lost. There are still powerfully developed paragraphs and pregnant dicta; but now Peirce rambles persistently, repeats himself, and employs expressions and arguments which, except for those fully acquainted with his thought, are obscure to a degree [Gallie (1952), 55-56].

There is a hint of this in Thomas A. Goudge's *The Thought of C. S. Peirce*. He finds that EG, at least as Peirce left it, seems far inferior to logical algebras, and he views the system as complicated and cumbersome:

One can hardly avoid the conclusion that in the end Peirce permitted his graphs to become . . . a "plaything". The fascination they exerted led to a steady increase in their internal complexity, without any corresponding increase in their positive results [Goudge (1950), 120].

Finally, Martin Gardner calls EG the most ambitious diagrammatic system yet attempted. His discussion, in *Logic Machines and Diagrams*, includes no examples of the graphs themselves, but is confined to general remarks. He says, for instance, that Peirce "held a greatly exaggerated notion of the value of his diagrams" (Gardner [1958], 58). Besides, Peirce's papers on the graphs

> are written in such an elliptic, involuted style that one is led to wonder if Peirce harbored unconscious compulsions toward cloudy writing that would enable him to complain later of his critics' inability to understand him. Add to this opaque style his use of scores of strange terms invented by himself and altered from time to time, and the lack of sufficient drawings to illustrate the meaning of these terms, and the task of comprehending his system becomes formidable indeed.[10]

Now with regard to this last point, Peirce was aware that EG involved a lot of new terminology, and an unusual notation; yet he continued to believe that it was essentially "very easy" (Ms 450, Ms 500). He gives this advice for approaching an exposition of the graphs:

> Let [the student] read these permissions and the commentary as he would listen to the rules of a new and intricate game, very closely attentive, but wide awake to the purpose of preparing for a lively and lightsome contest. The writer has a dear friend [William James?] of most active and agile intellect, and most spiritually minded withal. Moreover, he is a pragmatist of the strait sect himself. But when he comes to hear the writer lecture, he seats himself, contracts his brow, and evidently prepares himself for a tussle. The natural result is that he does not understand one word; while if he had made up his mind that understanding the doctrine was like stepping from a floating log into the water, he would have no more difficulty than the simple have, who always understand when they are not scared [Ms 280, pp. 23-24].

The next step is to "devote some hours daily for a week or two to practice with it" (4.617).

Peirce also believed that the graphs afforded the simplest available introduction to the logic of relatives. That this was important to him is clear from the following passage, written in 1898:

> When people ask me to prove a proposition in philosophy I am often obliged to reply that it is a corollary from the logic of relatives. Then certain men say, I should like exceedingly to look into this logic of relatives; you

[10] Gardner (1958), 55-56. Incidentally, Gardner's excellent little book (recently reissued as a paperback) contains some nice Peirce quotations that are not included in *CP*.

must write out an exposition of it. The next day I bring them a MS. But when they see that it is full of A, B, and C, they never look at it again. Such men—oh, well.[11]

The view that Peirce's later writings as a whole show evidence of encroaching senility cannot survive a serious study of those writings. The availability of the Peirce manuscripts on microfilm makes it much easier for anyone to engage in such a study. Nor is it true that Peirce's work on logic was finished by 1885. Recent studies by Fisch, Roberts, Turquette, and Zeman, for example, show that his later work must be considered significant (see the bibliography for these items).

As to the question of algebra versus diagram, Peirce recognized in 1902 that it was not logical graphs, but logical algebra which had chiefly been pursued (3.620). Still, in defining symbolic logic as deductive logic "treated by means of a special system of symbols" he was unwilling "to confine the symbols used to algebraic symbols, but [would] include some graphical symbols as well" (4.372). The algebraic notations continue to be popular—I like them myself—and it may be difficult to fit EG into such frameworks. But then, as Rulon Wells suggests, it might be rewarding to "shift attention to the problems of comparing frameworks" (Wells [1959], 210).

This book is intended to put the reader into a position to judge for himself about the value of the graphs. This involves, among other things, some understanding of the way EG fits into Peirce's philosophy. The next chapter, therefore, will show that EG is a development of several strains of thought that had occupied Peirce's attention since the late 1860's; Chapters 5, 6, and 7 examine similar and later connections. It should perhaps be pointed out that there are two or three passages in Chapter 2 which might be understood more easily after the exposition of EG has been given in full. As for the exposition itself, I hope it gives that simple introduction to the logic of relatives that the graphs were designed to provide.

11 The last sentence of the quotation, subsequently deleted by Peirce himself, originally read as follows: "Such men are intellectual *petit crevés*, nice to have around,---". Ms 437, p. 11. A slightly edited version of the passage appears in 1.629.

2

THE DEVELOPMENT OF EXISTENTIAL GRAPHS

2. 1 PEIRCE'S EARLY EMPHASIS ON DIAGRAMMATIC THINKING AND ANALYSIS

According to Peirce's own account, his first exposure to logic occurred when he was twelve or thirteen years old,[1] and he never lost interest in it after that. Perhaps as early as 1870, when his father defined mathematics as "the science which draws necessary conclusions" (4.229), Peirce thought of logic as a kind of analysis:

> In truth, no two things could be more directly opposite than the cast of mind of the logician and that of the mathematician . . . . The mathematician's interest in a reasoning is as a means of solving problems–both a problem that has come up and possible problems like it that have not yet come up. His whole endeavor is to find short cuts to the solution. The logician, on the other hand, is interested in picking a method to pieces and in finding what its essential ingredients are. He cares little how these may be put together to form an effective method, and still less for the solution of any particular problem. In short, logic is the theory of all reasoning, while mathematics is the practice of a particular kind of reasoning.[2]

Now 'the theory of all reasoning' is too broad for our study of Peirce's logic diagrams. Formal or deductive logic is our concern. Beginning with Peirce's characterization of mathematics as the practice of necessary reasoning, we concentrate on logic as the study or analysis of necessary reasoning.

Peirce's analysis of mathematical reasoning had convinced him, as early as 1869, that progress in mathematics, as in science, was tied to the use of observation. If it seems strange to speak of mathematics as a science for which observation is relevant, Peirce explains that it is "observation of artificial objects and of a highly recondite character"; indeed, "as the great mathe-

1 Peirce (1953), 27, 32 (a letter dated December 23, 1908). See also Murphey (1961), 17.

2 This is from pages 3-4 of the undated Ms 78 in which Peirce recalls how his father came to define mathematics as he did in 1870.

matician Gauss has declared–algebra is a science of the eye"(1.34). The use of letters in algebra, for example, provides a kind of diagram which can be experimented upon and observed.

Peirce thought that the use of such a notation in logic would be of value not only theoretically, but even practically, because of the economy and convenience of expression it would provide (3.45, 99). And in 1870, following "De Morgan's *Open Sesame*" (4.615), he proposed such a notation for the logic of relatives (3.45-149). Of course, the employment of such devices is at least as old as Aristotle, and even the ordinary syllogistic formula

All *M* is *P*
S is *M*
∴ *S* is *P*

"is really a diagram of the relations of *S*, *M*, and *P*. The fact that the middle term occurs in the two premisses is actually exhibited, and this must be done or the notation will be of no value" (3.363). But Peirce's extension of the use of such devices to the logic of relatives constituted a genuine advance.

Thus by 1870 Peirce had begun to develop an interest in diagrammatic thinking and the logic of relatives. These two interests, however, were not really combined or fused together until twelve years later. It is true that Peirce considered algebraic formulas to be diagrams of a sort; but it is also true that these formulas, unlike other diagrams, are not 'iconic'–that is, they do not resemble the objects or relationships they represent. Peirce took this to be a defect. But is was not untill 1882 that he attempted to construct a more iconic system of representation for the logic of relatives.

2.2 THE EARLIEST APPLICATIONS OF THE DIAGRAMMATIC METHOD TO THE LOGIC OF RELATIVES

For five years, from 1879 to 1884, Peirce taught logic at the Johns Hopkins University.[3] He was associated there with James J. Sylvester, who, with William K. Clifford, had begun (in 1877 or 1878) to use chemical diagrams to represent algebraic invariants (Murphey [1961], 196-197). It is very probable that this association was influential in Peirce's application of such diagrams to logic. Years later, in 1896, Peirce pointed out the connection between his entitative graphs (see below) and chemical diagrams, and he defined the term 'graph' as "Clifford's name for a diagram, consisting of spots and lines, in imitation of the chemical diagrams showing the constitution of compounds" (3.468, 469-470).

3 For an account of Peirce's stay at Johns Hopkins see Fisch and Cope (1952).

The first effects of this and earlier influences on Peirce are found in a letter he wrote to his student, O. H. Mitchell, on December 21, 1882. This letter contains perhaps the first attempt by anyone to apply diagrams to the logic of relatives in general. It is by no means the first application of diagrams to the purposes of logic; Leonhard Euler made use of circle diagrams to treat of the logic of classes (the traditional syllogistic) as early as 1761, and in 1880 John Venn introduced a system of interlocking circles for the same purpose.[4] But the traditional syllogism comprises only a small part of the logic of relatives in general.

Peirce's letter (Ms L 294) begins as follows:

The notation of the logic of relatives can be somewhat simplified by spreading the formulae over two dimensions. For instance suppose we write

to express the proposition that something is at once benefactor and lover of something. That is,

$$\Sigma_x \Sigma_y b_{xy} \ell_{xy} > 0.$$

Before we comment on the diagram itself, let us clarify Peirce's algebraic formulation of the proposition. The sigma, Σ, is Peirce's symbol for the existential quantifier, usually symbolized by ∃ in contemporary systems. Peirce used the pi, Π, for the universal quantifier. In contemporary systems, if a special symbol is used at all, it is usually the inverted A, that is, ∀. In Peirce's formula, the sign b_{xy} is a numerical coefficient whose value is 1 in case x is a benefactor of y, and O in the opposite case. A similar analysis is given the sign ℓ_{xy}. Peirce prefixed the quantifiers Σ and Π to such coefficients to form numerals, and then he formed statements from these numerals by adding the suffix >O. His reason for constructing the statements in this way was that "Any proposition whatever is equivalent to saying that some complexus of aggregates and products of such numerical coefficients is greater than zero".[5] In contemporary symbolism the quantifiers ∃ and ∀ are prefixed, not to numerical coefficients, but to propositional functions (or open sentences); and the result is not a numeral, but a statement or proposition. In *Principia*-style notation the proposition 'something is at once benefactor and lover of something' looks like this: (∃x)(∃y)[B(xy). L(xy)], where the dot represents conjunction. This formula can be read 'There is an individual x, and there is an individual y, such that x is benefactor of y and x is lover of y'.

4 A history of the use of logic diagrams is given in Gardner (1958).

5 3.351. See also 3.329 and the explanation in Berry (1952), 161-162.

Consider now the diagram itself. Note that the lines represent individuals–persons, in this case; and note that the lines, when simply drawn on the sheet, are to be read 'something' or 'someone'. The following diagrams illustrate certain other features of the 'system' Peirce had in mind. All four of them occur in his letter to Mitchell, and all four are given algebraic interpretations there. Figs. 1 and 2 further illustrate the use of the line as a sign of 'something', and Figs. 3 and 4 illustrate Peirce's notational device for expressing 'everything'. The interpretations given below are Peirce's.

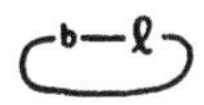

Fig. 1

Fig. 2

Fig. 3

Fig. 4

Fig. 1 means 'something is at once benefactor and loved of something, that is, something is benefactor of a lover of itself'. (Peirce's algebraic rendition of this expression is $\Sigma_x \Sigma_y b_{xy} \ell_{yx} > 0$; in *Principia* notation, $(\exists x)(\exists y)[B(xy).L(yx)]$.) Fig. 2 means 'something is a lover of itself' (that is, $\Sigma_x \ell_{xx} > 0$ or $(\exists x)[L(xx)]$). Fig. 3 means 'everything is a lover of itself' (that is, $\Pi_x \ell_{xx} > 0$ or $(\forall x)[L(xx)]$). In this system, to draw a bar through a line is to make it a sign of 'everything'. And the only difference between the diagrams of Figs. 2 and 3 is the bar that is drawn through the line of Fig. 3. Fig. 4 means 'everything is either a lover or a benefactor of everything' (that is, $\Pi_x \Pi_y (\ell_{xy} + b_{xy}) > 0$ or $(\forall x)(\forall y)[L(xy) \vee B(xy)]$–the wedge $\vee$ in the latter formula is used to represent non-exclusive alternation, while Peirce's symbol for this is simply an addition sign). This diagram affords a nice contrast with the diagram given at the top of page 18; the difference between them is that where the one diagram has 'everything', the other has 'something'.

Peirce does not clearly explain the shift from conjunction to alternation which occurs when bars are placed over the lines of the diagrams. The following quotation appears to be an attempt to do so:

> The order of attachment of like bonds is immaterial, that of unlike bonds is material. We can use shorter and straighter lines to represent later attached bonds. Thus
>
> 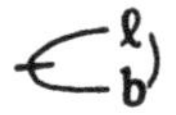
>
> will mean $\Sigma_y \Pi_x (\ell_{xy} + b_{xy}) > 0$, that is, there is something of which everything is either lover or benefactor; while
>
> 
>
> will mean $\Pi_x \Sigma_y (\ell_{xy} b_{xy}) > 0$ or everything is at once lover and benefactor of something or other.

We thus do away with the distinction of relative & non-relative operations, by discarding the latter altogether.

Two features of this notation–the use of the line to represent individuals and the use of the line, simply drawn, as a sign of 'something'–are basic in Peirce's final system of logic diagrams, the existential graphs. Yet it was not until 1896 that EG was discovered. Peirce also apparently noticed this anticipation of EG. In a manuscript written about 1906 he had this to say of the development of his system of graphs:

The system of expressing propositions which is called Existential Graphs was invented by me late in the year 1896, as an improvement upon another system published in the *Monist* for January 1897. But it is curious that 14 years previously, I had, but for one easy step, entered upon the system of Existential Graphs, reaching its threshold by a more direct way. The current of my investigations at that time swept me past the portal of this rich treasury of ideas. I must have seen that such a system of expression was possible, but I failed to appreciate its merits [Ms 498, pp. 1-2].

Peirce is almost certainly referring here to the 1882 letter to Mitchell, or to the reflections that led up to it. But what was that 'one easy step' to which he refers? Peirce does not say; but it is possible that he had in mind a device for expressing negation, since this was not provided for in the letter.

2.3 THE INFLUENCE OF KEMPE

In 1886 Alfred Bray Kempe published his "Memoir on the Theory of Mathematical Form" in the *Philosophical Transactions of the Royal Society of London.* This essay had a profound and lasting effect on Peirce. He must have taken it up almost immediately, for on January 17 of the following year he wrote a letter to Kempe about it which caused Kempe to reconsider and revise certain paragraphs of the Memoir.[6] Peirce described his experience with the Memoir in this way: "The paper is so difficult that I was at work upon it all day every day for about three weeks".[7] And while studying it, Peirce put together, for his own purposes, an index of terms of the Memoir (Ms 1170) and a brief list of definitions headed "Kempe translated into English" (Ms 715).

6 The revisions were published in Kempe (1887) and were prefaced by this comment: "An interesting letter of criticism from Professor C.S. Peirce on my recently published Memoir on the Theory of Mathematical Form has led me to reconsider certain paragraphs therein". The date of Peirce's letter is given in Kempe (1897), 453.

7 The old Houghton number for this manuscript was C,3(49). I have not been able to find its new number in Robin (1967).

Kempe states that his purpose, in the Memoir, is "to separate the necessary matter of exact or mathematical thought from the accidental clothing–geometrical, algebraical, logical, &c.–in which it is usually presented for consideration; and to indicate wherein consists the infinite variety which that necessary matter exhibits" (Kempe [1886], 2). In order to exhibit this mathematical form Kempe introduces a graphical notation of spots and lines (bonds) modeled, said Peirce (3.468), after the chemical diagrams which showed the constitution of compounds. The spots in Kempe's system represent 'units': the entities in terms of which the mind, in the process of reasoning, deals with the subject matter of thought. "These units come under consideration in a variety of garbs–as material objects, intervals or periods of time, processes of thought, points, lines, statements, relationships, arrangements, algebraical expressions, operators, operations, &c., &c" (Kempe [1886], 2). The lines, or bonds, connect certain of the spots–"one to one" (Kempe [1897], 457)–in order to divide the spots representing the units under consideration into two sets: one set whose units (or pluralities of units) differ from each other, and the other set whose units (or pluralities of units) do not. The lines, Kempe claims, are not used to represent "any relationship in the nature of a 'connexion', but simply to distinguish certain pairs of things from others".[8]

Peirce always referred to Kempe's essay in terms of the highest praise. For example, about 1905 he called the Memoir "the most solid piece of work upon any branch of the stecheology of relations that has ever been done", a work which "in addition to its intrinsic value, has that of taking us out of the logician's rut, and showing us how the mathematician conceives of logical objects".[9] But Peirce's admiration for the work did not prevent him from criticizing certain aspects of it, nor did it prevent him from doing things differently when constructing diagrammatic systems of his own.

Thus, in a manuscript dated January 15, 1889, entitled "Notes on Kempe's Paper on Mathematical Forms", the idea of representing individuals by the lines of the diagrams rather than by the spots again occurs to Peirce:

These ideas of Kempe simplified & combined with mine on the algebra of logic should give some general method in mathematics.

8 Kempe (1897), 458. See also 456.

9 5.505. See also 3.468, 3.601, and 6.174. By 'stecheology' (stecheotic, stoicheiology) Peirce meant "the general theory of the nature and meanings of signs, whether they be icons, indices, or symbols", 1.191, 4.9. This term, like many others in Peirce's specialized vocabulary, underwent some slight changes in meaning over the years. For our purposes, however, it is not necessary to trace the history of this term throughout the Peirce papers.

I note that lines may be treated as monads [This is not Kempe's procedure.] & so a new graph made & the question is whether there would be any advantage in this [Ms 714, p. 1].

As we have seen, this idea first occurred to Peirce in 1882, and it was to become one of the fundamental conventions of EG.

Other passages indicate that Peirce kept at the job of working out a diagrammatic treatment of logic during the years 1889 to 1896. Especially noteworthy is an article entitled "The Critic of Arguments", published in 1892. Consider in the first place the following remark:

I may mention that unpublished studies have shown me that a far more powerful method of diagrammatisation than algebra is possible, being an extension at once of algebra and of Clifford's method of graphs; but I am not in a situation to draw up a statement of my researches [3.418].

The collection of Peirce papers at Harvard University contains a number of undated manuscripts on logical graphs to which Peirce could here be referring. Consider in the second place Peirce's use of dashes in place of demonstrative terms and nouns to produce blank forms of propositions called 'rhemata'. If only one noun is erased, a non-relative rhema is produced, as '____ is mortal'. If two or more nouns are erased, the result is a relative rhema, such as '____ is bought by ____ from ____ for ____ ' (3.420). Consider in the third place the expressed analogy between logical compounds and chemical compounds:

A rhema is somewhat closely analogous to a chemical atom or radicle with unsaturated bonds. A non-relative rhema is like a univalent radicle; it has but one unsaturated bond. A relative rhema is like a multivalent radicle. The blanks of a rhema can only be filled by terms, or, what is the same thing, by "something which" (or the like) followed by a rhema; or, two can be filled together by means of "itself" or the like. So, in chemistry, unsaturated bonds can only be saturated by joining two of them, which will usually, though not necessarily, belong to different radicles. If two univalent radicles are united, the result is a saturated compound. So, two non-relative rhemas being joined give a complete proposition. Thus, to join "____ is mortal" and "____ is a man", we have "*X* is mortal and *X* is a man", or some man is mortal. So likewise, a saturated compound may result from joining two bonds of a bivalent radicle; and, in the same way, the two blanks of a dual rhema may be joined to make a complete proposition. Thus, "____ loves ____", "*X* loves *X*", or something loves itself [3.421].

This article is important for several reasons. It shows clearly that the parallel existing between these diagrams and chemical diagrams was strong in Peirce's mind; it shows Peirce again using a line to represent individuals; and it gives for the first time the kind of introduction to the terminology of the logic of relatives which occurs frequently in Peirce's later manuscripts, after he invented EG. The importance of this last point lies in the fact that Peirce here, before EG, introduces the logic of relatives in a manner that lends itself rather naturally to a diagrammatic system of representation.

As a final illustration of Peirce's continuing occupation with logic diagrams during this period, the following selection from an unpublished part of Peirce's *Grand Logic* is presented. The passage occurs in Chapter 7 of the book.

Such a proposition as, "Every mother loves some child of hers", is somewhat hard to put into a shape in which mere linkings take the place of inherences. But it is not uninstructive to consider how that could be done. Let us imagine an object which we may call "the character of being a mother". It is a thing, distinguishable from other things, which in every phenomenon of maternity is connected with that element of the phenomenon which presents a *possible* space-time continuity with several such phenomena to which it is related in the same way. That is the element we call the mother. This "character of being a mother" is conceived as a sort of badge which is attached to the mother. But it is not attached directly to the mother but only through the medium of some one of a series of objects

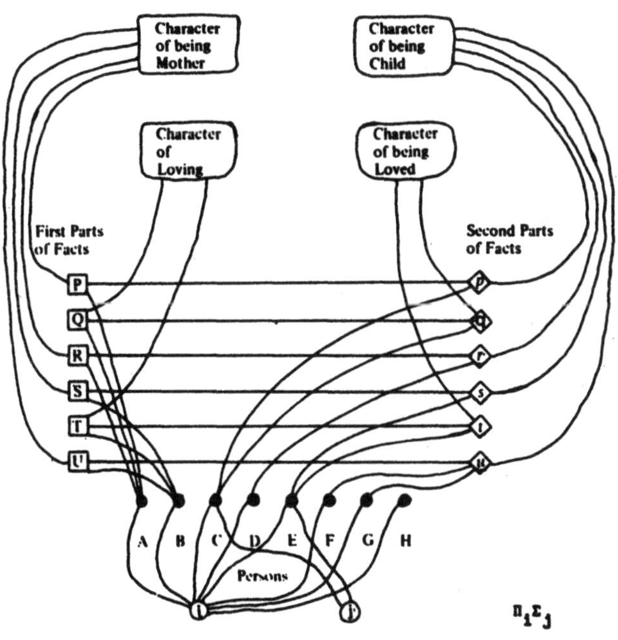

$\Pi_i \Sigma_j$

which we may call, in order to give them a general name, "first parts of facts". Every one of these objects is joined to another object of a collection which we call by the general name of "second parts of facts". We also have a special thing called "the character of being a child".

Let us imagine we have two other things called "the character of loving" and "the character of being loved". Finally, we imagine a quantity of objects called persons. Now look at the diagram.

It will be seen that person A is connected with half-fact P, which is conjugate to half-fact *p*. Half-fact P is joined to the character of being a mother, half-fact *p* to the character of being a child and also to person C. This is a way of representing that A is mother of C. So, through half-facts R and *r*, A is represented as mother of D; through half-facts S and *s*, B is represented as mother of E; and through half-facts U and *u*, B is represented as mother of F and G. Moreover, A is connected through half-fact Q with the character of loving; and the conjugate of Q is *q*, which is connected with the character of being loved and with person C. Thus, A is represented as loving C; and in like manner through half-facts T and *t*, B is represented as loving E. Thus, the diagram *exhibits* a state of things in which every mother loves one of her children. Being a mere icon, the diagram can do no more. But every person is connected with *i*; and C and E, the persons loved by their mothers, are connected with *j*. Against the diagram is written $\Pi_i \Sigma_j$ and this *asserts* that whatever person be taken, persons related to the rest as C and E can be found. Even with this attachment the diagram does not very distinctly assert the proposition intended; but it does serve to show very clearly that *inherence*, which it is the peculiar function of categorical propositions to express, is nothing but a special variety of connection [Ms 410, pp. 11-13].

As Peirce seems to admit, this is a rather complicated way to represent the proposition 'Every mother loves some child of hers'. Now Peirce's intention was, eventually, to construct a simple method of diagraming such propositions. He was soon to think that EG satisfied this condition. After the exposition of the Beta part of EG has been given, the existential graph of this same proposition will be presented, and it will be obvious then that the EG version is indeed simpler and clearer than the above diagram is (see section 3 of Chapter 4).

The above quotation from the *Grand Logic* of 1893 concludes with this comment by Peirce: "This diagram has been very obviously suggested by the ideas of Kempe's Memoir on Mathematical Form". Peirce apparently had in mind the use of spots to represent individuals, a device employed by Kempe. Peirce may also have been referring to his own use of lines as connections or relations between these spots; but, as was pointed out above, Kempe did not mean for the lines in his Memoir to be interpreted in this way. Rather, he meant to represent relations by appropriate spots, since these relations when considered by the mind are units, and units were to be represented by spots (Kempe [1897], 454, 457).

Peirce published his interpretation of Kempe's lines in the January issue of *The Monist* for 1897 (3.468, 479n.1), and Kempe replied in *The Monist* later that same year (Kempe [1897]). Peirce's last words on this minor controversy were never published, but it is worth noting that he did compose a "Reply to Mr. Kempe" in which he points out that "a 'connexion', 'mode of junction', or 'dyadic relation', is nothing more nor less than a class of pairs" (Ms 708, p. 2). This means that insofar as Kempe's lines serve to distinguish certain pairs from others, they do represent a dyadic relation of some sort.

2.4 THE FIRST SYSTEM: THE ENTITATIVE GRAPHS

Peirce's *Monist* article, just mentioned in the preceding paragraph, was entitled "The Logic of Relatives" (3.456-552). He intended it to be "a kind of popular exposition" of the work that was then being done in the field of the logic of relatives (3.456). He began by defining the term 'relation', and in order to do this adequately he introduced a system of logic diagrams.[10] Peirce later named this system the 'entitative graphs'.

The graphs are similar to those used by chemists to represent the constitution of matter. "A chemical atom is quite like a relative [term] in having a definite number of loose ends or 'unsaturated bonds', corresponding to the blanks of the relative" (3.469). In fact, the entitative graph of the proposition 'John gives John to John' (Fig. 5) corresponds closely to the chemical diagram for ammonia (Fig. 6):[11]

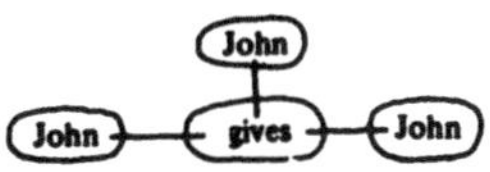

Fig. 5

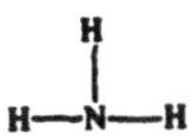

Fig. 6

Peirce pursues the analogy further than it seems useful to carry it here. What would be useful is to present a brief exposition of the basic conventions of this system of diagrams. This will now be done.

In the system of entitative graphs, to write a proposition (on whatever surface is being used for this purpose) is to assert it. Thus, Fig. 7 expresses the proposition 'Blue litmus paper is placed in acid'.

Fig. 7 Blue litmus paper is placed in acid.

To write two propositions on the same area is to assert an alternation whose component parts (alternants) are the two propositions. Thus, Fig. 8 expresses the alternative proposition 'Either blue litmus paper is placed in acid or the blue litmus paper will turn red'.

10 3.457 and 468ff. In fact, the diagrams are introduced as a means of achieving the third grade of clearness. An example of 'popular exposition' is this nice description of the logic of relatives as "nothing but formal logic generalised to the very tip-top" (3.473).

11 3.469. See also 4.561 n.1 (written about March, 1908).

Fig. 8 Blue litmus paper is placed in acid.
The blue litmus paper will turn red.

To encircle a proposition is to negate it. Thus, to express the proposition 'It is false that blue litmus paper is placed in acid' the graph of Fig. 9 is used.

Fig. 9
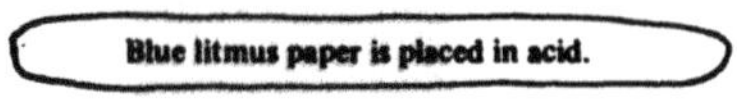

To assert a conditional proposition, such as 'If blue litmus paper is placed in acid then the blue litmus paper will turn red', the antecedent is placed within a circle, and the consequent is written on the sheet itself, as in Fig. 10.

Fig. 10
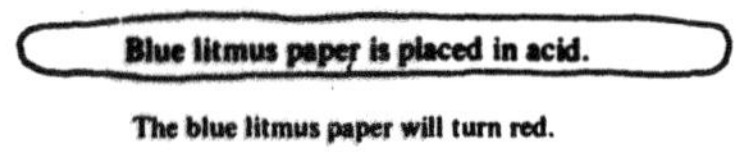

Finally, for the conjunctive proposition 'Blue litmus paper is placed in acid and the blue litmus paper will turn red', the graph of Fig. 11 is used.

Fig. 11

So far, of course, the symbolism is adequate only to the representation of non-relative propositions. That the system is able to do this, however, makes it something of an improvement on the notation of the 1882 letter to Mitchell (which could express *only* relative propositions); for with this much as a foundation, Peirce is able to provide a means for the expressing of relative propositions also. But before this can be done it is necessary to provide the system with a sign of the individual, a sign that can be taken to represent individual objects. Such a sign is provided by Peirce for the entitative graphs, and it turns out to be the sign he had used for this purpose in the letter to Mitchell and in the 1892 introduction to the logic of relatives: the dash —. In addition to this Peirce provides an interpretation of the line, based on its position in the graph, which indicates when the line is to be read as 'everything' and when it is to be read as 'something'. The rule is this: a line whose outermost part or extremity (least enclosed part) occurs unenclosed by circles or within an even number of circles is read 'all' or 'every'; and a line whose outermost part is enclosed by an odd number of circles is read 'some'.

With these conventions in mind, let us consider the following graphs:

Fig. 12

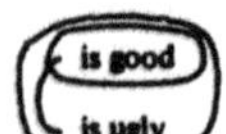

Fig. 13

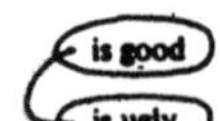

Fig. 14

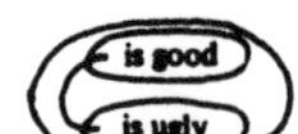

Fig. 15

In Fig. 12, the outermost part of the line in unenclosed; hence it is read 'all' or 'every'. The form of the proposition, considered apart from the line, is that of a conditional (see Fig. 10). Hence Fig. 12 asserts the proposition 'Everything, if it is good, is ugly', or, 'Everything good is ugly'.

In Fig. 13, the outermost part of the line is once (or oddly) enclosed; according to our conventions, it is to be read 'some' or 'something'. The form of the graph, considered apart from the line, is that of a conjunctive proposition (see Fig. 11). Hence Fig. 13 means 'Something is good and not ugly', or, 'Something good is not ugly'.

Similar analyses of Figs. 14 and 15 show that Fig. 14 means 'Nothing good is ugly' and Fig. 15 means 'Something good is ugly'.

2.5 THE SECOND SYSTEM: THE EXISTENTIAL GRAPHS

Peirce was not long satisfied with the system of entitative graphs. In fact, while he was reading the proof sheets of the *Monist* article in which the system appeared, a second system suggested itself to him.[12] He at once wrote out a full account of this new system and sent it off to Paul Carus, the editor of *The Monist*, hoping to have it published in a later issue. One of the more delightful accounts of the affair occurs in some papers on pragmatism written by Peirce in 1905 or 1906:

> The writer described a system of logical graphs, since named "Entitative Graphs", in [3.456-552]; but the ink was hardly dry on the sheets . . . when he discovered the far preferable system, on the whole, of Existential Graphs, which are merely entitative graphs turned inside out, and sent the gracious Editor a paper on the subject that could have been squeezed into a single number by simply excluding everything else. But the Editor feared that so swift the advances of exact logic seemed to be, that, before the types were half set up, the second system might be superseded. However, eight years have elapsed and one jot or one tittle has in no wise passed from the system.[13]

It is true that the basic conventions and rules of inference of EG remained the same over the years.[14] As might be expected, there were minor changes from time to time, regarding such things as the interpretation of the cut or enclosure and the symbolization of the pseudograph (this is taken up in Chapter 3). And there were proposed extensions to the system, such as the attempts to cover abstractions and modal logic, by means of special symbols

12 Ms 500; Ms Am 806*; Ms 498; Ms 280; Ms 513; and the letter to Woods begun October 4, 1913, Ms L 477.

13 Ms 280, pp. 21-22. The editor "preferred the old mumpsimus to my sumpsimus", Ms L 477.

14 See for example the entry in Peirce's *Logic Notebook* on p. 110r, dated June 14, 1898; 4.394-417 (1903); 4.530-572 (1906); and Ms 650 (July and August, 1910).

including the broken cut in 1903 and by means of the tinctures in 1906. But these topics are to be discussed in detail in later chapters.

Peirce said that existential graphs were "merely entitative graphs turned inside out", by which he meant that "Any entitative graph may be converted into the equivalent existential graph by, first, enclosing each spot separately and secondly enclosing the whole graph" (Ms 485, p. 1). The following brief exposition of EG will make this clear.

In EG, as in the system of entitative graphs, to write a proposition is to assert it. Hence the existential graph of the proposition 'Blue litmus paper is placed in acid', given in Fig. 16, is the same as the entitative graph of this proposition, given in Fig. 7.

Fig. 16 Blue litmus paper is placed in acid.

Furthermore, in EG as in the system of entitative graphs, to encircle a proposition is to negate it. Thus the proposition 'It is false that blue litmus paper is placed in acid' is expressed in the graph of Fig. 17, and Fig. 17 is the same as Fig. 9.

Fig. 17

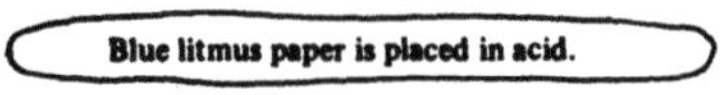

At this point the systems begin to diverge. For to write two propositions together in EG is to assert a conjunction, not an alternation. Fig. 18 is the EG expression of the conjunctive proposition 'Blue litmus paper is placed in acid and the blue litmus paper will turn red'.

Fig. 18 Blue litmus paper is placed in acid.
The blue litmus paper will turn red.

A comparison of Fig. 18 with Fig. 8 shows that the same device which expresses alternation in entitative graphs, expresses conjunction in EG.

Fig. 19 is the EG expression of the alternative proposition 'Either blue litmus paper is placed in acid or the blue litmus paper will turn red'.

Fig. 19

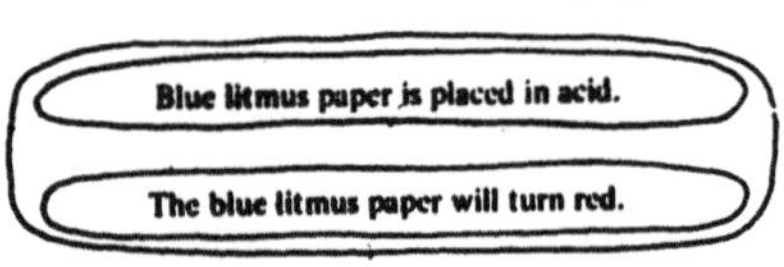

A comparison of this graph with the graph of Fig. 11 shows that the notation which expresses conjunction in entitative graphs, expresses alternation in EG.

Now consider the conditional proposition 'If blue litmus paper is placed in acid then the blue litmus paper will turn red'. This is expressed in EG by the graph of Fig. 20, below. The entitative graph of this same proposition is given in Fig. 10.

Fig. 20

We turn now to the EG representation of relative propositions. The device used in EG as the sign of the individual is like that used in the system of entitative graphs. In both cases Peirce used a line, but in EG he drew it quite heavily: ▬▬. Peirce called this the 'line of identity'. The position of this line also determines how it is to be read (whether by 'all' or by 'some'), but the rule for EG is the reverse of that used for the entitative graphs: a line of identity whose outermost part or extremity occurs unenclosed by circles or within an even number of circles is read 'some'; and a line whose outermost part is enclosed by an odd number of circles is read 'all' or 'every'.

Consider the following graphs:

Fig. 21

Fig. 22

Fig. 23

Fig. 24

In Fig. 21, the outermost extremity of the line is once (or oddly) enclosed, so that it is to be read 'all' or 'every'. The form of this graph, considered apart from the line of identity, is that of the conditional proposition. Hence Fig. 21 asserts the proposition 'Everything, if it is good, is ugly', or 'Everything good is ugly'.

In Fig. 22, the outermost extremity of the line of identity is unenclosed, and it must therefore be read 'some'. The form of the proposition, apart from the line, is that of a conjunction. Fig. 22 must mean 'Something is good and is not ugly', or 'Something good is not ugly'.

Similar analyses of Figs. 23 and 24 show that Fig. 23 means 'Nothing good is ugly' and Fig. 24 means 'Something good is ugly'. Thus Figs. 21-24 give us the four categorical propositions of traditional logic.

Why did Peirce say that EG is 'far preferable' to the entitative system? Because he thought that EG was simpler (Ms L 477) and much easier to use (Ms 513), and that the system of entitative graphs was unnatural in some of its basic conventions (Ms 484). The most unnatural feature of the system is its interpretation of the juxtaposition of two propositions as alternation. For since the writing down of a proposition asserts it, the writing down of two propositions naturally suggests asserting them both (as in EG). This feature renders the analysis of the conditional proposition less satisfactory in the entitative system than in EG. Finally, the greater simplicity of expression in EG shows up in connection with relative propositions. These things should be clear from the table in Appendix 2.

Peirce did not say just when he began calling his systems by the names 'entitative' and 'existential'. An early name for his second system was 'positive logical graphs'. This name occurs as the title of a manuscript, probably written in 1897, which begins:

The system of logical graphs here described is essentially the same as that which was sketched by me in a paper in the Monist for January 1897, except that the interpretation is nearly reversed. I shall distinguish the present system as positive, since it proceeds upon the principle that writing a proposition down asserts it [Ms 488].

As the system is explained, all the essentials of EG appear, such as the distinctive convention that writing down a number of propositions asserts them all.

The names 'entitative' and 'existential' occur in a letter Peirce wrote to William James on December 18, 1897 (Ms L 224). Peirce was telling James about his plans for the lectures he was to deliver at the Cambridge Conferences of 1898. 'Existential' first occurs in the *Logic Notebook* on June 9, 1898 (p. 102r), and frequently thereafter.

Why did Peirce choose these names? Because the "fundamental symbol" of the one system "expresses an entitative relation", and the fundamental symbol of the other "expresses the relation of existence" (Ms 485, p. 1). The second is easy enough to understand. In EG, to scribe or write something is to "aver that such a thing *exists*" (Ms 513), and is to claim that something having the character described exists in the universe which the sheet represents (Ms Am 806*). Not so easy is the meaning of the other name, which is that "being involves necessarily the truth of the description" (Ms 513).

3

ALPHA

The Alpha part of EG is the foundation of the entire system: the Beta part presupposes and builds upon Alpha, and Gamma presupposes and builds upon both Alpha and Beta. Alpha is concerned with the relationship between propositions considered as wholes. That is to say, it is a formulation of the propositional calculus, the logic of truth functions.

The presentation of Alpha will be given in two parts. In the first part I explain five conventions which are, in effect, formation rules for the system. In other words, these conventions amount to instructions in the reading and writing of the simplest kind of existential graphs. As you will see, there are only three basic symbols, or types of symbols, in Alpha: the sheet of assertion, the cut, and the graph. In the second part, rules of transformation for these graphs will be presented. These rules are instructions for operating with (or upon) the graphs, for transforming graphs already obtained into new, or other, graphs.

3.1 THE ALPHA CONVENTIONS

We are engaged in constructing a system of expression which will enable us to diagram, and then examine and experiment with, statements and inferences. The diagrams are to be two-dimensional figures; so we begin by providing for a two-dimensional surface on which these diagrams are to be drawn. This surface may in practice be a blackboard, or it may be a sheet of paper. In any case it will be called the 'sheet of assertion' (SA). The writing (or otherwise placing) of a graph-instance on SA will often be called 'scribing a graph', following the practice of Peirce (Ms 450, p. 8 verso). SA represents for us the 'universe of discourse', that is, the domain of objects to be talked about. Whatever we write upon it can be thought of as making the representation of the universe more determinate (Ms 455, pp. 2-3). Thus, suppose we write

A pear is ripe

on SA; this means that there is a pear in our universe, and it is ripe.

Of course, the universe of discourse represented by the sheet of assertion is not necessarily the real universe. For the logician as formal logician is not interested in whether or not the propositions he analyzes are factually true; he is, rather, concerned with certain relations that hold between them whether or not they are in fact true. The study of logic is often facilitated by analyses of propositions which are factually false, or even absurd. So the universe is an imaginary or fictitious universe, which becomes more determinate as the graphist (the one who scribes the diagrams on the sheet) proceeds with his business. In this way we account for the logical distinction between form and matter: the formal logician is interested only in the form of (that is, the truth relations between) propositions; the matter of the propositions (their actual truth or falsity) is not his concern (*LN* p. 103 r.-106 r.).

Now by 'graph' (or 'graph-instance') we mean any sign which expresses in a proposition any possible state of the universe. And according to Peirce, SA is a graph, even if nothing is written on it: "If the sheet be blank, this *blank*, whose existence consists in the absence of any scribed graph, is itself a graph" (4.397). And what does SA express? Whatever is taken for granted at the outset to be true of the universe of discourse. The sheet thus functions as a kind of all-purpose axiom, and we will understand the first convention (below) to be a statement of this axiom. See 7.22 for further discussion of this topic.

Peirce drew a distinction between the terms 'graph' and 'graph-instance', and he frequently illustrated it with the following example: The word 'the' will occur perhaps 20 times on an ordinary English page. Each of these occurrences or instances is a separate occurrence of the same word. "In the sense in which 'the' is *one* word only, no matter how many times it may occur, it is a *Type*",[1] that is, a general, a universal; each separate occurrence or instance of it is, in Peirce's terminology, a token. So also an instance of a graph is a token; the graph itself, expressed in the instance, is a type. Nevertheless, in this book, as in some Peirce papers such as Ms 280, the word 'graph' will frequently be used as an abbreviation of 'graph-instance'. Confusion is not likely to arise now that the distinction has been made.

We already have the first two conventions in hand: C1. *The sheet of assertion in all of its parts is a graph*. Not all of whatever surface is used for SA need be regarded as SA; the graphist may need some space (on his blackboard, say) for explanations or other comments. C2. *Whatever is scribed on the sheet of assertion is asserted to be true of the universe represented by that sheet*.

Our third convention explains what it means to scribe two or more graphs

1 Ms 498, p. 37. Cf. Ms 490, Ms 450, Ms L 231, and 4.537.

on SA. Suppose, for example, we have the following two graphs on the sheet:

A pear is ripe
The pulp of some oranges is red

We will understand this to mean that in our fictitious universe, it is true that there is a ripe pear, *and* that the pulp of some oranges is red. It is important to point out that the order or arrangement of graphs on the sheet has no significance. C3. *Graphs scribed on different parts of the sheet of assertion are all asserted to be true* (4.398). This makes juxtaposition the sign of conjunction. In another place Peirce puts it this way: "If different graphs are scribed on entirely different parts of the sheet of assertion, each shall have the same significance as if the other were not there" (Ms 450, p. 10).

By the term 'entire graph' is meant everything that is scribed on SA. A 'partial graph' is any graph scribed in the presence of, or along with, other graphs (Ms 450, p. 10). Since SA is a graph, the term 'total graph' is used to mean the entire graph together with SA itself.

It might be well to call attention here to a feature which is of importance later. To scribe

A pear is ripe

is to assert that something *exists* which is a pear and is ripe. And to scribe

The pulp of some oranges is red

is to assert that there *is* something which is red and is at the same time the pulp of some orange. In this system propositions asserted (or scribed on SA) have existential import: they imply the existence of whatever they describe—but existence, of course, only in the particular universe of discourse represented by the sheet.

C4 concerns the way in which EG is to express the conditional proposition. In a draft of one of the Lowell Lectures of 1903 Peirce used this diagram to distinguish and identify the parts of such a proposition, which he there called a 'consequence' (Ms 455, p. 6 verso):

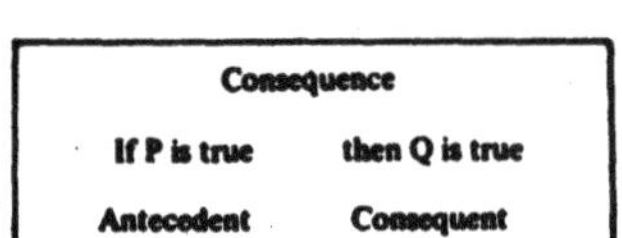

The kind of conditional in which Peirce was primarily interested is the truth-functional conditional, the material implication, the 'conditional *de inesse*'. The essential point is that this conditional does not imply that there is any real connection of any kind between the state of things *supposed* in the antecedent and the state of things *conditionally* asserted in the consequent. It asserts simply that *either* the antecedent is false *or* the consequent is true; "it limits itself to saying 'If you should find that P is true, then you may know that Q is true', never mind the why or wherefore" (Ms 455, p. 9). It follows from this that the only case in which this conditional proposition is false is when the antecedent is true *while* the consequent is false.

How shall we diagram 'If P then Q'? In order to assert it we must place it on SA. But since 'If P then Q' asserts neither P nor Q, we must be careful not to scribe them on SA.[2] We get what we want by means of what Peirce called a "scroll" – "two closed lines one inside the other" (Ms 450, p. 14), like this:

In one place Peirce gave the following instructions for drawing the scroll in four steps:[3]

First Second Third Fourth

You are to understand the ovals or lines of the scroll to be different from ordinary lines. We will "make believe" (Ms 455, p. 10) that they are cuts through the surface of SA, and that what is placed inside such a cut is severed from, or cut off from, the sheet itself (see Appendix 5, 'verso'). Suppose now that we place the graph Q in the innermost circle, and the graph P in the outermost compartment, obtaining this graph:

Note that we have succeeded in diagraming both P and Q, yet not on the surface of SA itself. And we agree to express in this way the conditional proposition *de inesse*: If P then Q. To express the proposition 'If some oranges have red pulp then naturalness is the last perfection of style', we scribe:

2 Here is a place where the analysis given by Peirce's first system of graphs, the entitative graphs, is less satisfactory than that afforded by EG. For in the earlier system the consequent of the conditional proposition is itself placed on the unenclosed sheet.
3 Ms 693, p. 292. Peirce was left-handed. The reverse direction might seem more natural to a right-handed person.

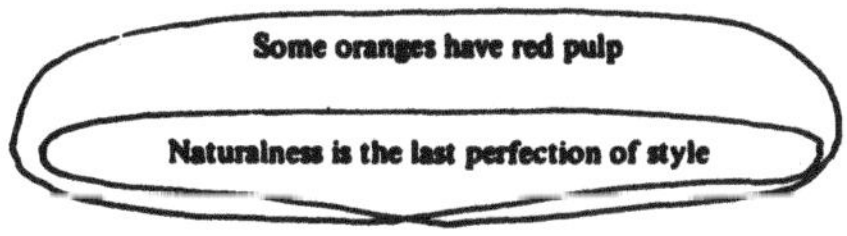

Here then is C4: *The scroll is the sign of a conditional proposition de inesse* (that is, of material implication) (Ms 450, p. 14).

It is important to note that the shape or size of the scroll has no significance. Thus, 'If P then Q' could just as well be diagramed in any of the following ways:

The essential feature of the scroll is that it consists of two closed lines one inside the other, and that we agree to place the antecedent in the outer compartment and the consequent in the inner one. We shall occasionally read this kind of graph 'P scrolls Q'.

Each of the lines that make up the scroll is called a 'cut';[4] and the space within a cut is called its 'close' (4.437) or 'area' (4.556). A cut taken together with its area and whatever is scribed on its area is called an 'enclosure' (4.399, 414(5), 437). A cut *per se*—the "self-returning finely drawn line" (4.414)—is not a graph, but an enclosure is.[5] The area on which a cut (or any graph) is made (or scribed) is called the 'place' of the cut (or graph). A cut that is not itself enclosed within another cut lies on SA; but no point within it lies on SA, since the cut separates its area from the sheet. Yet such an enclosure is said to lie on SA. The inner cut of a scroll may be called the 'loop', and its area the 'inner' or 'second' close or area (4.437). The 'outer' or 'first' close or area of a scroll is the area outside the loop but inside the outer cut. When the word 'close' is used of a scroll without qualification, it signifies the entire contents of the scroll, both the inner and outer areas.

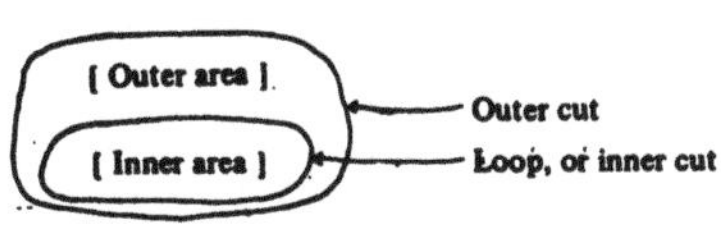

It was indicated in Chapter 2 that negation is expressed in EG by enclosing whatever we wish to deny in a single cut. At times, Peirce seemed to derive this convention from the nature of the conditional *de inesse*. For, as just remarked, the graph

4 In some places Peirce used the term 'sep', from *saepes*, a fence, 4.435.

5 4.399. This consideration is important for a proper understanding of the rules of erasure and deiteration, which are explained in section 3.2 below.

asserts that either the antecedent P is false, or the consequent Q is true; and "it all but follows that if the latter alternative be suppressed by scribing nothing but the antecedent, which may be any proposition, in an oval, that antecedent is thereby denied".[6] At other times his reasoning is based on the insight that a conditional with a false consequent expresses the denial of its antecedent.[7]

The filling up of any entire area with whatever writing material (ink, chalk, etc.) may be used shall be termed *obliterating* that area, and shall be understood to be an expression of the pseudograph on that area.

Corollary. Since an obliterated area may be made indefinitely small, a single cut will have the effect of denying the entire graph in its area. For to say that if a given proposition is true, everything is true, is equivalent to denying that proposition.[8]

Peirce sometimes drew a cut entirely filled in, or blackened–

and used this 'blot' to symbolize the pseudograph.[9] At other times he simply wrote out an example of a proposition which is always false: 'Every proposition is true', or 'No proposition is false' (see 4.452). But he also employed the empty cut–i.e., the cut with a blank area–for this purpose (4.467). And if this is done, it follows once more that a cut precisely denies its contents. To see this, consider what it takes to diagram the proposition 'It is false that it rains'. This is accomplished by the following scroll

6 4.564. In Ms 650, p. 20, Peirce says "Before I had the concept of a cut, I had that of two cuts, which I drew at one continuous movement as shown": 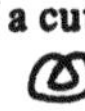

7 For more on this method of expressing negation see Appendix 1, item G-1885-3, (6).

8 4.402. Cf. 4.455. One disclaimer to this piece of reasoning was published in a footnote to 4.564 (on p. 452); it is from "Copy T" of Ms S 30, which was an early draft of the October 1906 article "Prolegomena to an Apology for Pragmaticism", *not* a manuscript "designed to follow" Prolegomena, as the editors of *CP* state (on page 453). In the final draft of Prolegomena, Peirce reinstated his earlier view (4.567), and he continued afterwards to support the interpretation given in C5. See, for example, 4.617 (1908), Ms 650 (July and August 1910), and Ms L477 (November 1913).

9 Ms 450 (pp. 23-24) and 4.455. He called this symbol the 'pseudograph' because, strictly speaking, it is not a graph; for, being absurd, it is an expression of an impossible state of things, whereas a graph is the propositional expression of a possible state of the universe. He called it 'the' pseudograph because all such propositions (all absurd propositions) are equivalent. Ms 450, p. 22; 4.395. In one account Peirce gave the pseudo-

since we have here a conditional with a false consequent (the empty cut is the pseudograph). Now (and we anticipate part of our later exposition) it is a basic rule of transformation that two cuts, one within the other but with nothing scribed in the outer area, can be drawn on any area or removed from any area where it occurs. Hence the above graph is equivalent to the following graph:

And hence the cut denies its contents.

The above is summarized in C5: *The empty cut is the pseudograph; and the cut precisely denies its contents.*

Cuts cannot intersect one another in EG, but (as the scroll illustrates) they may be made within other cuts. This produces what Peirce called 'nests' of cuts. A nice definition of this term occurs in an unpublished manuscript (you should recall that the area of a cut is the space it encloses, while the place of a cut or graph is the area on which it is made or scribed):

> If there be a collection, i.e., a definite, and individual plural of cuts, of which one is placed in the sheet of assertion, and another encloses no cut at all, while every other cut of the series has the area of another cut of this collection for its place, and has its area for the place of still a third cut of this collection, then I call that collection a *nest*, and the areas of its different cuts *its successive areas*, and I number them ordinally from the sheet of assertion as *origin*, or zero, with an increase of unity for each passage across a cut of the nest inwards that one can imagine some insect to make if it never passes out of an area that it has once entered. For example, in Fig. 1 there are five nests as follows:

Fig. 1

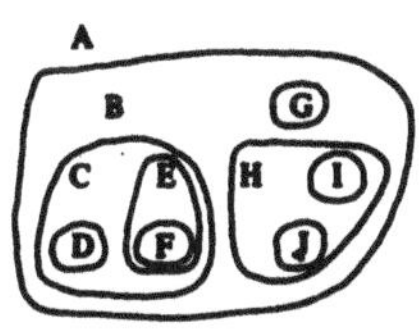

1. One of 5 areas, or 4 cuts; A-B-C-E-F,

2., 3., 4. Three of 4 areas or 3 cuts each; A-B-C-D, A-B-H-I, A-B-H-J,

5. One of 3 areas, or 2 cuts; A-B-G.[10]

---

graph the distinction of being the only graph which cannot be erased from SA, 4.415(2).

10 Ms 650. In Ms 693, p. 292, Peirce writes "A *nest* is any series of cuts each enclosing the next one".

An area is said to be oddly enclosed if it is enclosed by an odd number of cuts; it is evenly enclosed if it is enclosed by an even number of cuts or by no cuts at all. This terminology applies also to graphs scribed on such areas. Thus, in Fig. 1, the letters A, C, F, G, and H are evenly enclosed; the others are oddly enclosed. Not only the letters, but any part of the graph which is itself a graph can be described as evenly or oddly enclosed. For example, (F) and (G) are oddly enclosed, and the scroll (E (F)) is evenly enclosed.

It is convenient to inject some order into the areas of graphs. For any graph P, let '{P}' denote the place of P. And let the relation symbol $\supseteq$ be defined as follows: {B} *is enclosed by every cut that encloses* {A} *if and only if* {A} $\supseteq$ {B}. The sign $\supseteq$ may be read 'contains'.[11] Examination of Fig. 1 in the light of this definition shows that {A} (namely, SA) contains every area in the graph, and {B} contains every area except {A}. It is also clear that the relation holds only between areas which belong to the same nest; for example, neither {E} $\supseteq$ {G} nor {G} $\supseteq$ {E} is true.[12]

According to C2 and C5, whatever is scribed on SA is asserted, and whatever is enclosed in a single cut is denied. Fig. 2

P

Fig. 2

Fig. 3

Fig. 4

therefore means 'P is true', and Fig. 3 means 'P is false'. In Fig. 4 the graph P is enclosed in two cuts, and must therefore mean 'It is false that P is false'–that is, 'P is true'. Further examples might suggest that whatever is enclosed in even cuts is asserted, while whatever is enclosed in odd cuts is denied. But this is too easy; it would mean that Fig. 1 asserts A, C, and G (among other things), and denies B, D, and I (etc.). Peirce had something else in mind.

11 The idea for the present treatment is due to Thomas Lee Schafter, a former student of mine (Schafter, 1968). The relation $\supseteq$ gives a partial ordering of the areas of existential graphs. This means that for arbitrary graphs P, Q, and R, the following propositions hold:

(a) {P} $\supseteq$ {P}. (Containment is reflexive.)

(b) If {P} $\supseteq$ {Q} and {Q} $\supseteq$ {P}, then {P} = {Q}, i.e., P and Q are scribed on the same area. (Containment is antisymmetric.)

(c) If {P} $\supseteq$ {Q} and {Q} $\supseteq$ {R}, then {P} $\supseteq$ {R}. (Containment is transitive.)

The reader may verify these for himself. Peirce used the notion of areas "immediately or mediately contained within" other areas when stating the rules of iteration and deiteration in Ms 490, but he did not there treat containment as a reflexive relation.

12 If, on the contrary, for arbitrary areas {P} and {Q}, either {P} $\supseteq$ {Q} or {Q} $\supseteq$ {P}, in addition to (a), (b), and (c) of note 11 above, the areas of graphs would constitute an ordered system.

Consider Fig. 5. According to C4 this means 'If P then Q'.

Fig. 5

Notice that we do *not* read it: 'Q is true and P is false', even though Q is evenly enclosed and P is oddly enclosed. Rather, the proposition Q in the inner close is regarded as posterior to or consequent upon the proposition P in the outer close. In other words, we read the graph from the outside (or least enclosed part) and we proceed inwardly, a method to which Peirce gave the name 'endoporeutic'.[13]

By applying this endoporeutic method but with our attention focused on C5, we find a different but equivalent reading of Fig. 5. We begin with the outer cut, which is to be read, 'It is false that ....' But what is being denied? Why, whatever is contained in the cut–in this case, it is the graph of Fig. 6. Now Fig. 6, according to C3 and C5, may be read in this way: 'P is true *and* Q is false'.

Fig. 6

Fig. 5 is the precise denial of this, and must therefore mean: 'It is false that: P is true and Q is false'. And this is the same as 'If P then Q'.

There is a third way to read Fig. 5. The truth-functional conditional, we said earlier, does not imply that there is any real connection between the state of things supposed in the antecedent and the state of things conditionally asserted in the consequent; the conditional asserts simply that either the antecedent is false or the consequent is true. Thus Fig. 5 may be read 'Either P is false or Q is true'.

Consider now Fig. 7. How is this to be read? You can answer this by comparing Fig. 7 with Fig. 5.

Fig. 7

The only difference is that where Fig. 5 has the graph of P, Fig. 7 has the graph of not-P. Since Fig. 5 is read 'If P then Q', Fig. 7 can be read 'If not-P then Q'. Again, since Fig. 5 may be read 'It is false, that P is true and Q is

13 "The interpretation of existential graphs is *endoporeutic*, that is proceeds inwardly; so that a nest sucks the meaning from without inwards unto its centre, as a sponge absorbs water.... If anybody were to find fault with the system for expressing truth endoporeutically, –as if we opened a closet to whisper it in, instead of speaking out and ever further out, I should be disposed to admit that it is a poetical fault. But I had difficult enough conditions to fulfil in constructing the system, without considering purely esthetical points in its essential features" (Ms 650, pp. 18-19). In Ms L 477 (dated 1913) Peirce uses the term 'endogenous' for the method of interpretation which begins from the outside and proceeds inward. An 'exogenous' interpretation would begin in the inside and proceed outward.

false', Fig. 7 can be read 'It is false, that P is false and Q is false'. Finally, since Fig. 5 may be read 'Either P is false or Q is true', Fig. 7 can be read 'Either P is true or Q is true'. These interpretations are summarized and related to other notations in Appendix 2.

How shall we read Fig. 1? The basic structure of the entire graph is that of a conjunction: A is true *and* the enclosure is true. But the enclosure can be read in several different (though equivalent) ways. Perhaps the simplest reading is to view it as a scroll whose antecedent is the conjunction (because juxtaposed on the same area) of B, (G) and (H (I) (J)). The graph as a whole can then be read as follows: A is true; and *if* B is true, G is false, and at least one of H, not-I, and not-J is false, *then* C is true, D is false, and E implies F (see 7.21).

3.2 THE ALPHA RULES OF TRANSFORMATION

> The object of reasoning is to find out, from the consideration of what we already know, something else which we do not know. Consequently, reasoning is good if it be such as to give a true conclusion from true premisses, and not otherwise [5.365].

This was written in 1877, and repeated 26 years later in an exposition of EG (4.476ff). Since the purpose of EG is to analyze and therefore represent reasoning (more on this in Chapter 7) the system must be provided with rules of inference of some kind; and if these rules are to represent good reasoning, they must be valid–i.e., they must be such as never to transform a true premiss into a false conclusion. A few remarks about this will be made as we present the rules, but a full scale proof of the validity of the rules and of the consistency of the system is reserved for Appendix 4. Even for present purposes, however, it is convenient to have the method of truth-table analysis available. Let 1 and 2 represent the truth-values truth and falsehood, respectively. By 'value of an area' is meant the value of the conjunction (juxtaposition) of all the graphs scribed on that area. And (by C3) this value is calculated by the rule that a conjunction has the value 1 if each conjunct has the value 1. This means that a single 2 on an area is sufficient to give the area the value 2. We indicate the value of an area by placing a 1 or a 2 inside square brackets on that area. The value of an enclosure, indicated by a 1 or a 2 placed just outside its cut, is 2 if the value of its area is 1, and it is 1 if the value of its area is 2 (C5). To illustrate the procedure, we show that 'P scrolls Q' has the value 2 when P = 1 and Q = 2:

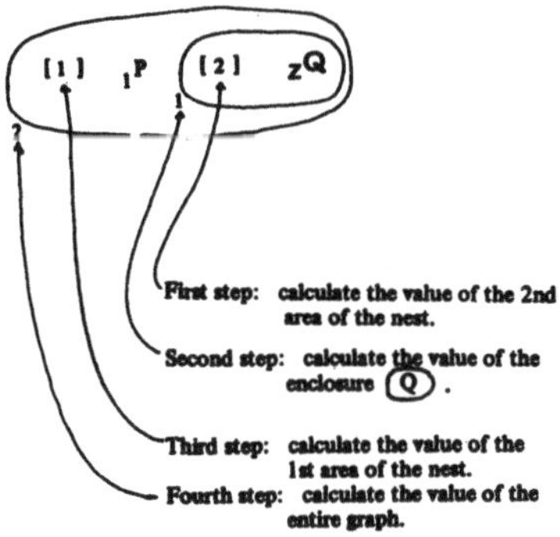

R1. The rule of erasure. *Any evenly enclosed graph may be erased.* Suppose that the graph of Fig. 1 below is scribed on SA. This graph means 'P is true, and Q scrolls R'.

Fig. 1

Fig. 2

P

Fig. 3

Fig. 4

According to R1, this entire graph can be removed, leaving the blank SA. The result of this transformation can hardly result in a false assertion; for to assert nothing (or to assert the blank SA, which by C1 has the value 1) is not to assert something false (Ms 650, p. 13). R1 permits the transformation of Fig. 1 into Fig. 2 and also into Fig. 3, for if a conjunction is true, then each of its conjuncts is true. The rule of erasure also justifies the transformation of Fig. 2 into Fig. 4, for this transformation consists in the erasure of R, which is twice (and therefore evenly) enclosed in Fig. 2. But what does Fig. 4 mean? And will it be true if Fig. 2 is true?

Compare Fig. 4 with Fig. 2. Fig. 2 means 'If Q is true then R is true'. But, as was pointed out earlier, this graph can also be read 'It is false that Q is true and (R) is true'. By analogy, therefore, Fig. 4 must mean: 'It is false that Q is true and ◯ is true'. But ◯ is the pseudograph, and has (always) the value false, or 2; hence the value of the first area of Fig. 4 is 2, and the value of the entire graph is 1, or true. Fig. 4 then, is true regardless of the value of Q, and regardless of the value of Fig. 2; and it follows from this (trivially) that the inference of Fig. 4 from Fig. 2 by means of R1 is valid.

R2. The rule of insertion. *Any graph may be scribed on any oddly enclosed area.* Consider Fig. 5. According to R2, Fig. 5 can be derived from Fig. 2;

Fig. 2

Fig. 5

for the transformation of Fig. 2 into Fig. 5 consists in the inserting of the graph S into an area once (and therefore oddly) enclosed. Suppose now that Fig. 2 has the value 1; then the first area of the nest is the conjunction of Q and Ⓡ – and must have the value 2. The addition of S to that first area of Fig. 2 cannot change the value of that area from 2 to 1, regardless of the value of S. Hence Fig. 5 must be true.

R3. The rule of iteration. *If a graph P occurs on SA or in a nest of cuts, it may be scribed on any area not part of P, which is contained by {P}.*[14] That is to say, any graph which occurs on some area may be scribed again on that area, or on any area enclosed by additional cuts. In Fig. 1 the graph P occurs on SA. According to R3, P can be scribed again on SA, as in

Fig. 1

Fig. 6

Fig. 7

Figs. 6 and 7; on the once enclosed area as in Fig. 8; on the twice enclosed area as in Fig. 9; or on both enclosed areas as in Fig. 10.

Fig. 8

Fig. 9

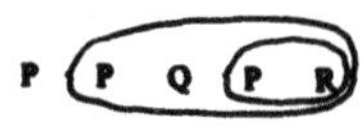

Fig. 10

In Fig. 1, Q occurs once enclosed. Hence, by iteration, Q can be scribed again on the same area, as in Fig. 11; or on the twice enclosed area as in

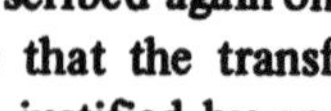

Fig. 11

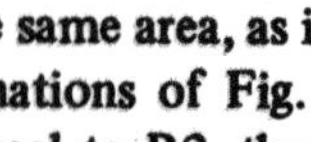

Fig. 12

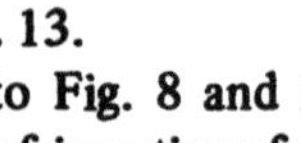

Fig. 13

Fig. 12. But this rule does not justify the scribing of Q on SA itself, since SA is not contained by {Q}. In Fig. 1 the graph R occurs twice enclosed. By R3 it can be scribed again on the same area, as in Fig. 13.

Notice that the transformations of Fig. 1 into Fig. 8 and into Fig. 11 could be justified by an appeal to R2, the rule of insertion; for they both consist in scribing a graph on an oddly enclosed area. The distinctive thing about the rule of iteration is that it permits the insertion of a graph into an evenly enclosed area when the conditions for its use are satisfied. Peirce likened the use of this rule to

14 Keep in mind that {P} denotes the place of P. It might be useful at this time to reread (a) and (c) of note 11, as well as the definition of $\supseteq$ in the text over the footnote. Peirce stated this rule in several different ways. One of the simplest statements, which unfortunately does not make it clear that a graph may be iterated on the area of its original occurrence, is in Ms 650: "*If you have a right to scribe a graph on the place of a cut, you have a right to scribe it on its area*; whence it follows that any graph that is scribed on any area may be iterated, or repeated, on any higher-numbered area of the same nest".

. . . a march to a band of music, where every other step only is regulated by the arsis or beat of the music, while the alternate steps go on of themselves. For it is only the iteration into an evenly-enclosed area that depends upon the outer occurrence of the iterated graph, the iteration into an oddly enclosed area being justified by your right to insert whatever graph you please into such an area, without being strengthened or confirmed in the least by the previous occurrence of the graph on an evenly-enclosed area [Ms 650].

The clause 'not part of P' in R3 is intended to prevent such transformations as that from Fig. 14 into Fig. 15, since for P = 1,

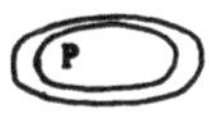
Fig. 14

Fig. 15

Fig. 14 is true but Fig. 15 is false. Fig. 15, in fact, is a contradiction and is false for all values of P. The clause blocks this inference by making clear that the iteration of  onto its own area is illegitimate.[15]

R4. The rule of deiteration. *Any graph whose occurrence could be the result of iteration may be erased.* It is not necessary that the graph in question actually be the result of R3, but it is necessary that the conditions for this are in fact satisfied. Hence, in order to erase any graph by R4, there must be at least two instances of that graph in a given nest of cuts; and it is the more-times-enclosed instance that may be erased (if the instances occur on the same area, all but one of them may be erased by R4). For example, R4 justifies the transformation of Fig. 11 into Fig. 1, since this transformation consists in the erasure of one occurrence of Q from an area containing two occurrences of Q. By R4, Fig. 8 can be transformed into Fig. 1, since this transformation consists in the erasure of the graph P from an area once enclosed, but it leaves another occurrence of P outside this area, that is, enclosed by fewer cuts than one. A valid application of deiteration always leaves another instance of the graph which was removed, and it leaves that other instance either on the same area as that occupied by the removed instance or on an area enclosed by fewer cuts than that occupied by the removed instance.

15 E. James Crombie trapped this fallacy, brought it to my attention, and helped me tame it. The particular illustration of it in Fig. 14 was suggested to me by my student David Stephenson. The spurious use of R3 which leads from Fig. 14 to Fig. 15 will also lead from any theorem to a contradiction of the form of Fig. 15, and the use of R5 will make the contradiction even more apparent. To see this, substitute any theorem, say 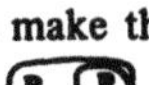 for P in Fig. 14, and perform the spurious inference.

Notice that although the transformations of Fig. 13 into Fig. 1, of Fig. 12 into Fig. 1, and of Fig. 9 into Fig. 1 could be justified by an appeal to R4, this is not necessary. Since they all consist in the erasing of a graph from an evenly enclosed area, they can be justified by R1, the rule of erasure. The distinctive thing about the rule of deiteration is that it permits the erasure of a graph that is oddly enclosed, when the conditions for its use are satisfied.

Both R3 and R4 are truth-preserving. Consider for example the transformation of Fig. 1 into Fig. 9 by iteration. Suppose the premiss, Fig. 1, has the

Fig. 1

Fig. 8

Fig. 9

value 1. Then both P and 'Q scrolls R' have the value 1, and the first area of the nest in Fig. 1 has the value 2.

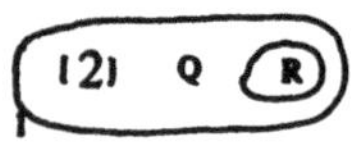

This means that either Q = 2 or Ⓡ = 2. Of course, if Q = 2, the first area of the nest in Fig. 9 has the value 2, from which it follows that Fig. 9 has the value 1. So suppose that Q = 1. Now the addition of P to the second area of the nest of Fig. 1 changes things only if P = 2, for this would give the value 1 to (P R), and the scroll in Fig. 9 would take the value 2. But the assumption was that P = 1; hence, if Fig. 1 is true, so is Fig. 9, and the inference in question is valid. A similar chain of reasoning will show that the inference from Fig. 8 to Fig. 1 by deiteration is also valid. For if Fig. 8, the premiss, is presumed to have the value 1, then P = 1, and the first area of the nest in Fig. 8 must have the value 2 independently of P–and this guarantees that the first area of the nest in Fig. 1 has the value 2, so that Fig. 1 as a whole must be true.

In order to state R5 we need to define a new term. A scroll with nothing on its first area (its outer close or area) is called a double cut. R5.The rule of the double cut. *The double cut may be inserted around or removed (where it occurs) from any graph on any area*. Since any part of SA is a graph (according to C1), R5 enables us to scribe a double cut anywhere on the blank SA. In addition, this rule permits the transformations of Fig. 1 into Figs. 16, 17, and 18,

Fig. 1

Fig. 16

Fig. 17

Fig. 18

and it permits the reverse transformations back into Fig. 1. All four of these graphs are equivalent. It is left to the reader to perform the truth-value analyses which show these transformations to be valid.

For convenience of reference the EG conventions and rules of transformation are collected together in Appendix 3. Here, for purposes of illustration, we present two proofs in EG .

3.3 FURTHER ILLUSTRATIONS

(1) The rule of *modus ponens.* This rule of inference may be stated as follows: From P and 'If P then Q' we may infer Q. In terms of the graphs, we must justify the transformation of Fig. 1 into Fig. 2:

Fig. 1

Q

Fig. 2

There are in fact two ways to do this. One way is to perform the transformations on the premisses themselves, so that when the inference is completed only the conclusion remains on SA. The other, and more instructive way, is to perform the actual transformations on iterated instances of the premisses, so that when the inference is completed the steps leading to the conclusion remain on SA as a kind of record. The first method is preferred by students writing examinations. We employ the second method here. The steps in our proofs are numbered, and to the right of each step we give the justification for it.

1.  Premiss.

2.  From 1 by R3.

3.  2, 1 by R4: deiterating the once enclosed P of step 2, since P occurs unenclosed in step 1.

4. Q 3, by R5.

In future proofs we will not bother to write out iterations which are as obvious as that of step 2 above, leaving it to the reader to supply them whenever he considers it necessary for clarity. Such an abbreviated proof of *modus ponens*, for example, might consist of steps 1, 3, and 4.

(2) The self-distributive law of material implication. This is stated in *Principia* notation as follows:

$[P \supset [Q \supset R]] \supset [[P \supset Q] \supset [P \supset R]]$. In EG it looks like this:

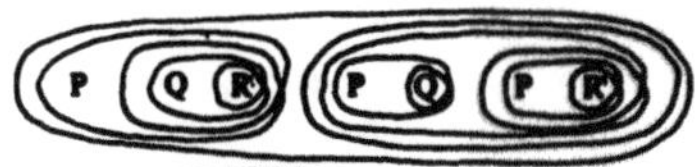

1.  R5.

2. 

1, by R2.

3. 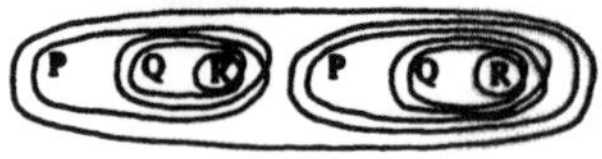

2, by R3.

4. 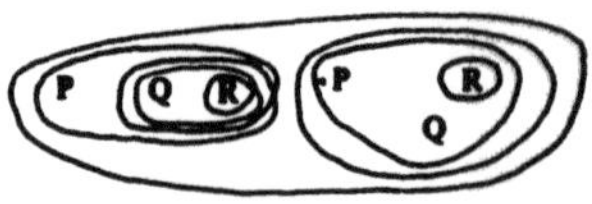

3, by R5.

5. 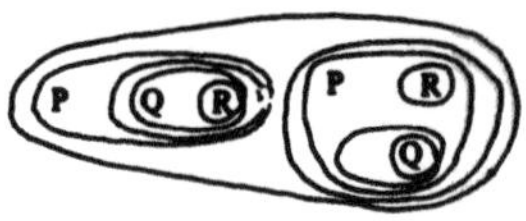

4, by R5.

6. 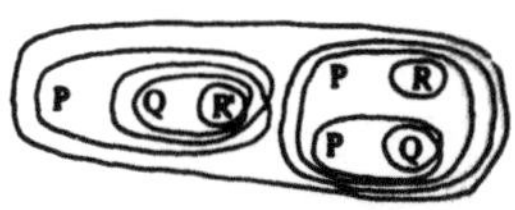

5, by R3.

7. 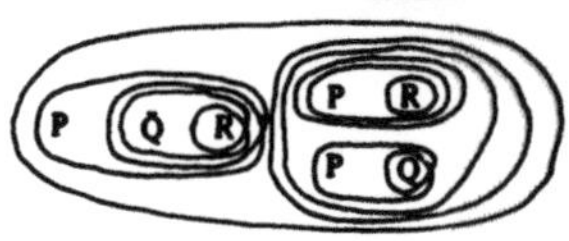

6, by R5.

4

BETA

In one manuscript Peirce says that the Beta part of EG is distinctive in that it takes account of individual identity and individual existence (Ms 462, pp. 8, 34). In another place he says that it is by means of the Beta part that we are able to express categorical propositions (Ms 450, p. 20). Beta is, in fact, a treatment of the functional or predicate calculus, the logic of quantification.

The Beta conventions introduce two new symbols: the line of identity and the spots. The numbering of the conventions continues that of Chapter 3.

4.1 THE BETA CONVENTIONS

The first Beta convention (the sixth in our list) provides a way for us to denote a single individual object; it is done by scribing a heavy dot or a heavy line on SA, like this • or this —. Peirce argued that since the universe of discourse (represented by SA) is a collection of individuals, any decidedly marked point of the sheet should be taken to stand for a single individual existing in that universe. Hence, — will mean 'something exists' (Ms 455, p. 21). Furthermore, since the 'fact' that something exists is taken for granted at the outset (*Ibid.*), since this fact is one of the "indemonstrable implications" of the blank SA (4.567, and see the discussion of C3 in Chapter 3), Peirce treated the unattached line (or the dot) as an axiom and permitted it to be scribed on SA (4.417, 567). We capture all of this in C6: *The scribing of a heavy dot or unattached line on the sheet of assertion denotes the existence of a single, individual (but otherwise undesignated) object in the universe of discourse. And it is always permitted to scribe such a dot or line on the sheet.* So far, EG appears to be a two-axiom system: there is the blank SA of C1, and the unattached heavy line of C6. In Appendix 4 it is shown that no more axioms are needed.

Suppose now that two heavy lines were placed on the sheet: ═ . How are we to read this? Are we to say that there are two different individuals in the universe, or do the two lines refer to the same individual? C3 stated that graphs scribed on different parts of SA are all asserted to be true; hence we must read this graph 'Something exists and something exists'. But our discussion of C3 also pointed out that graphs scribed on different parts of the sheet are to have the same meaning as if each stood alone; hence, in the absence of further information, we must treat the two 'somethings' as different. This will mean, in practice, that a graph refers to as many individuals as it contains *disconnected* heavy lines (Ms 481).

If we scribe

—has red pulp

on SA we mean to assert 'There exists something in the universe which has red pulp'. If we scribe

—has red pulp
—is an orange

we mean to assert that something is an orange and that something has red pulp; but nothing is said about the relation between these two somethings. In order to express that the very same thing is an orange and has red pulp, the two lines are joined as in the following graph:

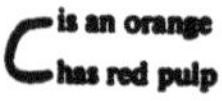

As Peirce puts it, we simply agree that the heavily marked line, all of whose points are *ipso facto* heavily marked and therefore denote individuals, shall be a graph asserting the identity of all the individuals denoted by its points (Ms 455, p. 23). Here is C7: *A heavy line, called a line of identity, shall be a graph asserting the numerical identity of the individuals denoted by its two extremities.*

In most of the graphs we have used so far there occur English words as well as special graphical symbols. Thus, to express the proposition 'There is an orange which has red pulp' we scribed this graph:

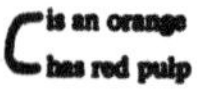

Here the English words 'is an orange' and 'has red pulp' have not been reduced to symbols; these 'unanalyzed' parts of the graph, the English phrases, are called 'spots' or 'rhemata' (see 2.3 above and 7.121 below). The places about each spot to which a line of identity can be attached are called the 'hooks' of the spot. In the above graph each spot has one hook; but we will in the following often be concerned with spots having more than one hook. Thus the following graph requires two hooks:

—is an opponent of—

The next graph has two spots, each of which has two hooks:

—is between—and—

It is possible to scribe the same graph with only one spot which takes three hooks, as follows:

Regardless of the number of its hooks, every spot that is scribed must be provided with a line of identity (or a heavy dot) at each hook (4.474); otherwise the spot is not a graph (4.441, 560). This will be discussed in Chapter 7 below, but an important result may be mentioned in passing: formulas with free individual variables cannot be expressed in EG. It is interesting that in one early exposition of the graphs, Ms 493 ("Principle 5"), Peirce did present a means for showing, should it be necessary, that "a complete assertion is not intended". The later and definitive expositions of EG, however, permit no graph to have an empty hook.

Suppose now we allow the line of identity to branch, as in the following graphs:

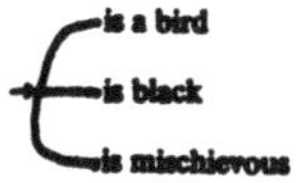

This means that there is a black bird which is mischievous. We could consider the heavy lines as a single line of identity with three extremities; but Peirce usually preferred to consider the figure as three lines of identity which have a point in common, namely, the point marked by the arrow. And the totality of all the lines of identity that join one another he called a 'ligature'. I prefer the former terminology, but both will occur in the book. The point in the above graph from which the three lines of identity can be taken to proceed, has the force of the conjunction 'and' (Ms 455, p. 24); hence the graph could be read 'There is something which is a bird *and* is black *and* is mischievous'.

There is practically no limit to the number of ways a line of identity can be made to branch. Consider, for example, this description of Aristotle:

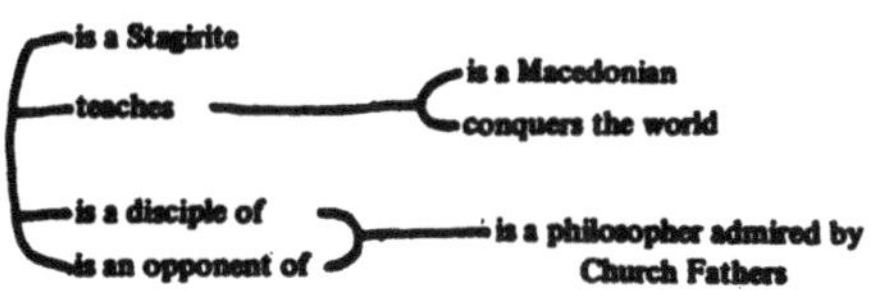

'There is a Stagirite who teaches a Macedonian conqueror of the world and who is at once a disciple and an opponent of a philosopher admired by Fathers of the Church' (Ms 450, p. 18).

You are now prepared for C8: *A branching line of identity with n number of branches will be used to express the identity of the n individuals denoted by its n extremities.*

It remains to explain how the line of identity is to be used and interpreted when it occurs in graphs involving cuts. Fig. 1 means 'There exists

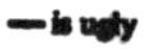

Fig. 1

Fig. 2

something which is ugly', or 'Something is ugly'. If we place a cut around this graph we obtain Fig. 2. Since by C5 the cut precisely denies its contents, this graph means 'It is false that something is ugly', which is to say 'Nothing is ugly'.

Note that according to this method of interpretation, a line of identity on SA stands for *some* individual object; but a line of identity once enclosed can be read as standing for *every* or *any* individual of the kind mentioned. Words like 'some' and 'any', or 'something' and 'anything', tell us "how to proceed in order to experience the object intended"; and such words Peirce called 'selective' pronouns (Ms 484, § 3). He gives the following explanation:

> The words *some* and *any* conceive an individual to be selected. *Some* means that a suitable individual is to be chosen by the speaker, or person interested in sustaining the truth of the proposition, while *any* means that the choice of the individual may be left to the listener, or to a person who might be hostile or sceptical to the proposition [Ms 503, p. 3].

Consider now Fig. 3. Here we have a line of identity crossing a cut.

Fig. 3

How is this to be read? Notice first that there are, in a sense, two lines of identity: one outside the cut, and one inside the cut.[1] These two lines have a point in common (the point on the cut) so that they form a ligature; hence, by C8, they must be taken to denote the same individual. If the line were broken, as in Fig. 4, it would be easy to interpret. According

Fig. 4

to C3, Fig. 4 means 'Something exists *and* it is false that something is ugly'. Now the essential thing about Fig. 3 is that it identifies these two some-

1 Except for a brief period in the spring of 1906 (4.579; see Chapter 6 below, section 1), Peirce allowed no graph to rest partly on one area and partly on another (4.405-407, 414(6), 416(11); Ms L 231). Nevertheless, it is quite natural and inconsequential to speak of a line of identity as crossing a cut, as Peirce did himself. In some cases he was careful to state that the part of the line that crosses is not a graph (4.449, especially the last sentence; 4.474(4), 4.561); but in others he did not bother to explain (4.458,459).

things; so Fig. 3 means 'Something exists and it is false that *this same something* is ugly'. That is, 'Something is not ugly'.

Place a cut around the graph of Fig. 3, and you get Fig. 5

Fig. 5

which reads: 'It is false that something is not ugly', that is, 'Everything is ugly', or 'If anything is, it is ugly' (cf. Ms 481, Fig. 34).

We have just said, in connection with Figs. 1 and 2, that according to our method of interpretation a line of identity on SA represents *some* individual object; and a line of identity once enclosed represents *every* individual in the universe. Does this 'rule of interpretation' hold true for Figs. 3 and 5 also? The answer to this is not immediate; for although Fig. 3 is read '*some*-thing is not ugly', only part of that line of identity lies on the sheet of assertion itself. And although Fig. 5 is read '*every*-thing is ugly', only part of that line of identity is once enclosed.

The 'principle secret' of interpretation (as Peirce phrased it in Ms 454, p. 18) is this: we are to consider a line of identity to be as much enclosed as its least enclosed (that is, its outermost) part (cf. 4.387); and if this part is on SA it denotes *something suitably chosen*, but if its outermost part is once enclosed it denotes *anything you please* (cf. 4.458). Thus, since the outermost part of the ligature in Fig. 3 lies on SA, the ligature refers to some individual suitably chosen; and since the outermost part of the ligature in Fig. 5 is once enclosed, that ligature refers to any individual in the universe.

We extend this principle at once to read as follows: A line of identity is as much enclosed as its least enclosed part; and if this outermost part is *evenly* enclosed it refers to a suitably chosen individual, while if this outermost part is *oddly* enclosed the line refers to any individual taken at pleasure. Now in logic systems generally, a symbol which denotes a suitably chosen individual, an existing individual, is called an existential quantifier; and a symbol which refers to any individual at all, taken arbitrarily from the universe of discourse, is called a universal quantifier. The evenly enclosed line, then, is an existential quantifier, and the oddly enclosed line is a universal quantifier. One word of caution: as we have already seen, graphs containing lines enclosed by cuts can be read in more than one way; the principle given here supposes that the line of identity is read *before* the cut which encloses it is read.

Here are some more examples. Fig. 6 means 'Something good is ugly'.

Fig. 6

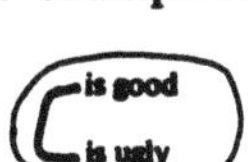

Fig. 7

Fig. 8

Fig. 9

Place a cut around this graph and Fig. 7 results: 'It is false that something good is ugly', that is, 'Nothing good is ugly'. A comparison with Fig. 3 shows that Fig. 8 reads 'Something good is not ugly'. And if we place a cut around this last graph, we obtain Fig. 9, which means 'It is false that some good thing is not ugly', or 'Everything good is ugly'. In Figs. 6 to 9 we have expressed the four categorical propositions of traditional logic.

Consider now Figs. 10 and 11. Fig. 10 is quite similar to Fig. 9 (the A proposition of traditional logic), with one major difference: here there are two disconnected lines of identity to account for. According to C7, this means that Fig. 10 refers to two individuals. And according to the method of

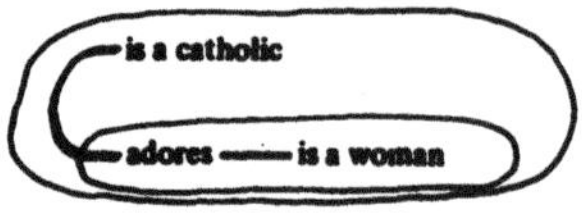

Fig. 10

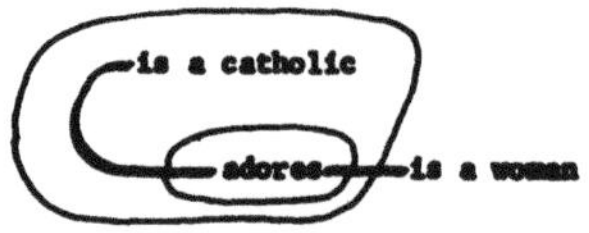

Fig. 11

interpretation discussed above, the line whose outer extremity is once (and therefore oddly) enclosed refers to every individual described as a catholic; and the line whose outer extremity is twice (and therefore evenly) enclosed refers to some individual described as an adored woman. (Note that all of this second line is evenly enclosed; hence the outer extremity of the line must be evenly enclosed.) Fig. 10 can therefore be read as follows: 'Every catholic adores some woman'.

What about Fig. 11? This differs from Fig. 8 (the 0 proposition of traditional logic) by containing two disconnected ligatures. Consider first the graph which would result from Fig. 11 if the outer cut were removed:

Fig. 12

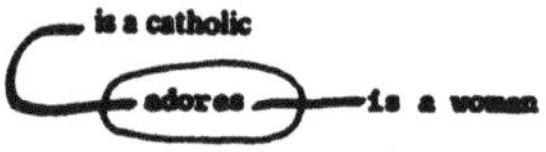

This means 'There are two individuals; one is a catholic and the other is a woman; and it is false that the catholic adores that woman'. In other words, 'There is some catholic who does not adore some particular woman'. Now if we restore the outer cut to obtain Fig. 11 again, and begin our interpretation (endoporeutically) at the outermost part of the graph, it reads as follows: 'There is some woman *and* it is false [this is the force of the restored outer cut] that some catholic does not adore this woman'. The second part of this reading, 'It is false that some catholic does not adore this woman', is a denial of an 0 proposition; but this is equivalent to the assertion of the A proposition 'Every catholic adores this woman'. Therefore Fig. 11 may be read 'There is some woman, and every catholic adores her', or 'There is a woman whom every catholic adores'.

Figs. 10 and 11 are thus by no means equivalent, even though the only difference is that a line of identity with its spot which is evenly enclosed in both graphs is evenly enclosed in two cuts in Fig. 10, and is evenly enclosed in no cuts in Fig. 11. Fig. 10 asserts that every catholic adores some woman or other; but Fig. 11 asserts that all catholics adore the same woman.[2]

The important point is the order in which the individuals are selected. It is one thing, for instance, to claim that for any individual you show me, I can find some other individual loved by the first one; it is quite another thing, and considerably more risky, to claim that I can point out some individual now who is loved by anyone you may select in the future. According to Peirce, the difficulty of representing the logic of relatives graphically lies entirely in the circumstance that it is necessary to distinguish between these two propositions (Ms 481). And in EG, this distinction is made—that is, the order of selection is determined—by the endoporeutic method of interpretation: First all unenclosed lines are read, then those that are once enclosed, then those that are twice enclosed, and so on.

One case involving a line crossing a cut remains to be discussed. It is the case in which the line of identity passes entirely through an empty cut, as in Fig. 13:

Fig. 13

This device signifies the non-identity of the individuals denoted by the extremities of the ligature: 'There are two objects such that no third object is identical to both'. Fig. 14 expresses that 'Three individuals are not all

Fig. 14

Fig. 15

identical' (4.469), and Fig. 15 says 'Everything is other than something' (Ms 277, p. 178). In Fig. 16, where X and Y are the individuals denoted by the points on the inner cut touched by the two lines, the meaning is 'If X is

Fig. 16

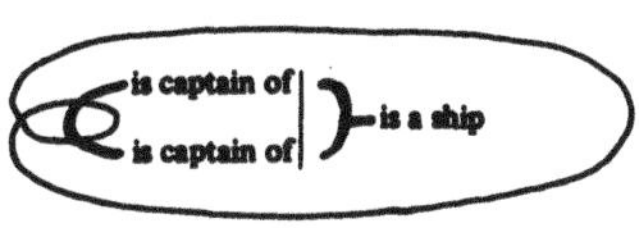

Fig. 17

the sun and Y is the sun, X and Y are identical' (4.407). Fig. 17 asserts that

2 Ms 504, p. 3. Peirce's reading of Fig. 10 is "Take anybody, A. Then somebody, B, can be found such that, either A is not a catholic or A adores B and B is a woman". And for Fig. 11 he gives "Somebody B can be found such that taking anybody, A, B is a woman and either A is not a catholic or A adores B". Cf. 4.452.

no ship has two captains. If a graph such as this last one is not quite obvious to you yet, try reading the outer cut first, as 'It is false that . . . .'

What if the ends of the line in Fig. 13 are joined? The resulting graph, Fig. 18, must mean 'Something exists which is not identical to itself'.

Fig. 18

Fig. 19

Without the cut we have Fig. 19, 'Some individual is identical to itself'.

So far all the graphs we have used involving lines of identity have been such that the lines were either entirely enclosed in one area or crossed a cut; we have not yet considered cases in which the line terminates on the cut. There are essentially two such cases; one in which the line which terminates on the cut lies outside the cut (i.e., on the place of the cut); and one in which it lies inside that cut (on the area of the cut). Since the line may be evenly or oddly enclosed for each case, there are four possibilities to account for:

Fig. 1

Fig. 2

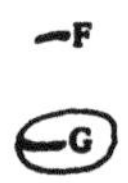

Fig. 3

Fig. 4

In both Figs. 1 and 2 the line which terminates on a cut lies outside that cut; it is evenly enclosed in Fig. 1, and oddly enclosed in Fig. 2. In Figs. 3 and 4 the line terminating on a cut lies inside that particular cut; in Fig. 3 it is oddly enclosed, and in Fig. 4 it is evenly enclosed.

Only one new convention is required for the interpretation of these graphs. C9. *Points on a cut shall be considered to lie outside the area of that cut* (the cut, says Peirce in 4.501, "is outside its own close"). The connection of a point on a cut with any other point within or without the cut is to be interpreted as if the point on the cut were outside of and away from the cut.

To make this quite clear, let us consider how we are to interpret Figs. 1 to 4 above. According to C9, the graph of Fig. 1 is equivalent to that of Fig. 1a; both mean 'Something is F and nothing is G'. Similarly, Fig. 2 is equivalent to Fig. 2a; they mean 'Nothing is F or something is G' (or, what is the same thing, 'If something is F then something is G'). Fig. 3 is equivalent to Fig. 3a; these graphs mean 'Something is F and something is not G'. And

Fig. 1a

Fig. 2a

Fig. 3a

Fig. 4a

Figs. 4 and 4a are equivalent; they mean 'Either nothing is F or everything is G' (or 'If something is F then everything is G').

Graphs may occur whose lines of identity cannot be drawn without their crossing one another (see 4.460). To indicate that such lines are to be distinguished (for their intersection would identify all the individuals denoted) Peirce introduced several devices at different times. The earliest is illustrated in Fig. 5. It was called a 'frog' in Ms 494 (p. 1) and was used there and in Ms 450 (p. 25) to distinguish crossing lines from each other; but in Ms 492 it was used to indicate that the lines intersect and are one. The parallel cross lines of Fig. 6 were introduced in 4.462 and 4.460 (Ms

Fig. 5

Fig. 6

Fig. 7

492), and are illustrated below in 5.13. Fig. 7 illustrates what Peirce called a 'bridge' in 4.561. We may imagine it to be a bit of paper ribbon, with one line passing under it and the other line upon it (Ms 455, p. 24). The bridge is illustrated in Fig. 8 below and in 5.11.

Still another way to avoid the crossing of lines is to replace some of them at each of their attachments by a capital letter, using a different capital letter for each different line. These letters function as names of the individuals denoted by the lines they replace (4.460-461, 561), and they also serve as quantifiers according to this rule: the area of its outermost occurrence determines how a letter is to be read, whether as 'some' or as 'any'. It is perhaps because of this last function that the letters are called 'selectives'; this is suggested by the discussion of selective pronouns immediately following the statement of C8 above.

To illustrate the use of selectives and the bridge, I scribe the graphs of Figs. 8 and 9, each expressing the proposition 'Some woman loves all of her children'. Fig. 8, with the lines of identity, employs the bridge; Fig. 9 employs selectives.

Fig. 8

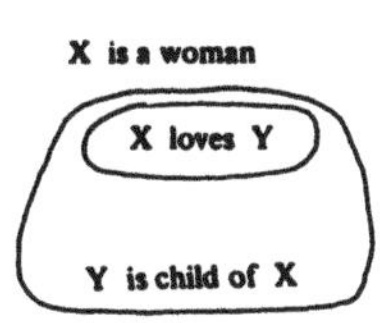

Fig. 9

Peirce also employed selectives when introducing new symbols, as in 5.11 below. The use of letters for names facilitates the reading of symbols given in a list. But however used, the selectives were considered as abbreviations only, and Peirce gave good reasons for rejecting their steady employment (4.473; 4.561n.1 and the continuation of the note in Ms 300).

4.2 THE BETA RULES OF TRANSFORMATION

The Beta rules are, essentially, extensions of the five Alpha rules. Since this is so, and since the Alpha rules are thus part of the Beta rules, we preserve the names and numbering given in Chapter 3.

R1. The rule of erasure. *Any evenly enclosed graph and any evenly enclosed portion of a line of identity may be erased.* By this rule Fig. 1, 'Something is F but not G', can be transformed into Fig. 2, 'Something is not G'. And Fig. 3 can be transformed into either Fig. 4 or Fig. 5 (and some other graphs as well). Fig. 5, incidentally, means 'It is false, that something is F and the pseudograph is true'.

Fig. 1

Fig. 2

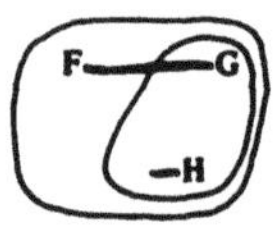

Fig. 3

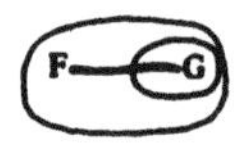

Fig. 4

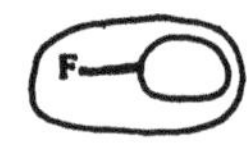

Fig. 5

The new clause in R1, about portions of a line of identity, permits the transformations of Fig. 1 into Figs. 6 and 7.

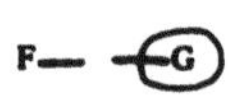

Fig. 6

Fig. 7

Figs. 6 and 7 are equivalent according to C9. Fig. 4 means 'Whatever is F is also G'. By R1 it can be transformed into Figs. 8 and 9, which mean 'If something is F then something is G'. That Figs. 8 and 9 are equivalent follows from the endoporeutic method of interpretation, according to which it is the outermost extremity of a line of identity that determines how it is to be read.

Fig. 8

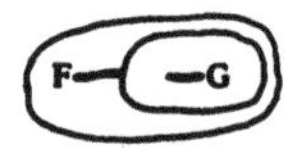

Fig. 9

R2. The rule of insertion. *Any graph may be scribed on any oddly enclosed area, and two lines of identity (or portions of lines) oddly enclosed on the same area, may be joined.* R2 justifies the transformation of Fig. 1 (above) into Fig. 10, 'Something is F, and either that same something is not G or else nothing is H'. And Fig. 4 can be transformed into Fig. 11, 'Either whatever is F is G, or else nothing is H'.

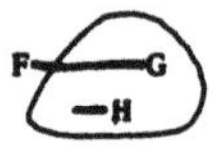

Fig. 10

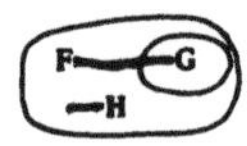

Fig. 11

The new clause in R2 permits the transformation of Fig. 12, 'Either nothing is F or nothing is G', into Fig. 13, 'Nothing is both F and G'.

Fig. 12

Fig. 13

It also permits the transformations of Fig. 14 into Fig. 1, and of Fig. 15 into Fig. 4.

Fig. 14

Fig. 15

For if it is true that something is F but nothing is G (Fig. 14), then it is true that something is F and not G (Fig. 1). And if it is the case that if something is F then everything is G (Fig. 15), then it is the case that whatever is F is G (Fig. 4).

R3. The rule of iteration. *If a graph P occurs on SA or in a nest of cuts, it may be scribed on any area not part of P, which is contained by {P}. Consequently,* (a) *a branch with a loose end may be added to any line of identity, provided that no crossing of cuts results from this addition;*[3] (b) *any loose end of a ligature may be extended inwards through cuts;* (c) *any ligature thus extended may be joined to the corresponding ligature of an iterated instance of a graph; and* (d) *a cycle may be formed by joining, by inward extensions, the two loose ends that are the innermost parts of a ligature.*

Whether this rule is applied to graphs involving lines of identity or not, it requires that the iterated graph be an exact replica of some other graph already scribed. And when lines of identity are involved, those in the iterated graph must correspond exactly to those in the original graph: they must join the corresponding hooks of the corresponding spots.

Clause (a) of the rule permits the transformation of Fig. 1 into Fig. 2, and that of Fig. 3 into Fig. 4. These transformations amount to an extension of a line on its own area, and (together with those permitted by clause (a) in R4, below) they indicate that the shape or size of the line has no significance.

Fig. 1

Fig. 2

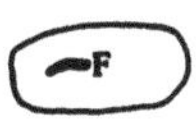

Fig. 3

Fig. 4

3 Peirce states this part of the rule (and its converse, R4(a)) in 4.505, but without the restriction. Christopher Gray pointed out to me that the restriction has to be stated, and I subsequently found an account in which Peirce made it explicit (Ms 490, the passage omitted at the ellipsis marked in 4.581).

And since, according to C9, points on a cut are considered to lie outside and away from the cut itself, clause (a) of R3 also permits the transformations of Fig. 5 into Fig. 6, and of Fig. 7 into Fig. 8.

Fig. 5

Fig. 6

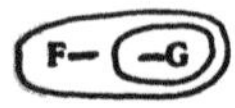

Fig. 7

Fig. 8

Clause (b) permits the transformation of Fig. 9 into Fig. 10 (the P represents any graph).

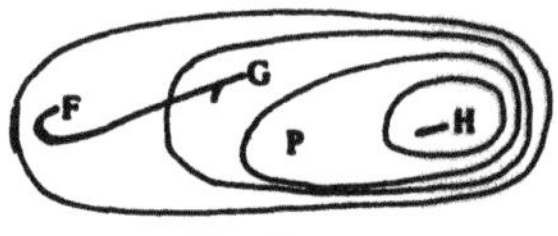

Fig. 9

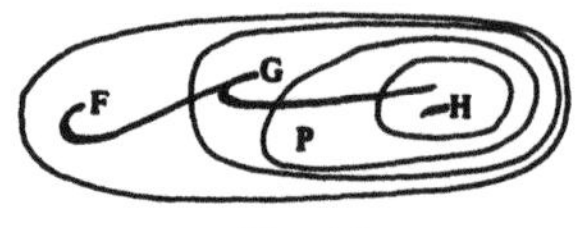

Fig. 10

And if R3 has already justified the transformations of Fig. 11 into Fig. 12,

Fig. 11

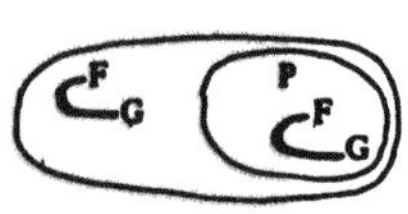

Fig. 12

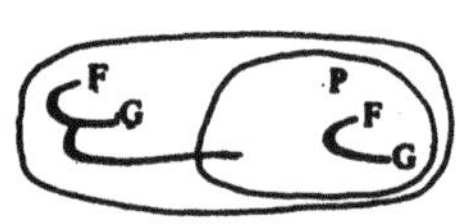

Fig. 13

and of Fig. 12 into Fig. 13, then clause (c) permits the transformation of

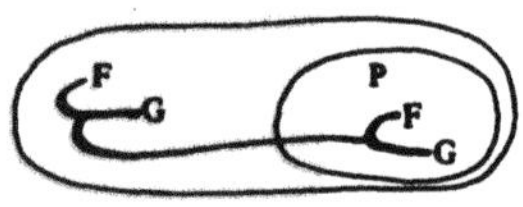

Fig. 14

Fig. 13 into Fig. 14. Notice that the extension of the original line, as in Fig. 13, does not affect the way in which the outer extremity of the line is enclosed, so that the interpretation of the line remains unchanged. And the joining of the two lines, as in Fig. 14, adds nothing essentially new: for the identification of the individuals denoted by the two lines is understood from the first–since the iterated instance is an iteration of the original line.

Clause (d), finally, permits the transformation of Fig. 15 into Fig. 16:

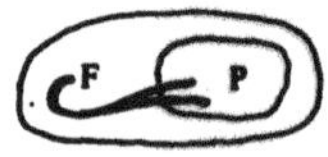

Fig. 15

Fig. 16

R4. The rule of deiteration. *Any graph whose occurrence could be the result of iteration may be erased.* Consequently, (a) *a branch with a loose end may be retracted into any line of identity, provided that no crossing of cuts occurs in the retraction*; (b) *any loose end of a ligature may be re-*

tracted outwards through cuts; and (c) *any cyclical part of a ligature may be cut at its inmost part.*

In general, R4 enables us to get back to the condition of things before R3 was employed. Clause (a) of R4, thus, permits the transformations of Fig. 2 (under R3 above) into Fig. 1, Fig. 4 into Fig. 3, Fig. 6 into Fig. 5, and Fig. 8 into Fig. 7. And (b) of R4 permits the transformation of Fig. 10 into Fig. 9.

Note that R4 has no clause corresponding to (c) of R3; yet by R4 we can obtain Fig. 11 (above) from Fig. 14, thus in effect reversing the transformations justified by R3 (c). The main clause of R4 permits the transformation of Fig. 14 into Fig. 17, and then clauses (b) and (a) transform this graph into Fig. 11. (The alert reader will recognize that the inference from Fig. 14 to Fig. 17 could also have been justified by R1; the illustration here is meant to show how R4 can be employed, and R1 cannot justify deiteration of graphs from oddly enclosed areas as R4 can.)

Fig. 17

Finally, clause (c) of R4 permits the transformation of Fig. 18 into Fig. 19.

Fig. 18

Fig. 19

We come at last to R5, the rule of the double cut. We want to extend this rule to permit the transformations from Fig. 20 into Fig. 21, and back again to Fig. 20.

Fig. 20

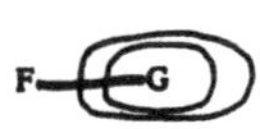

Fig. 21

Both these graphs mean the same thing, viz., 'Something is both F and G'. Nevertheless, these transformations require an extension of the rule, since, strictly speaking, a double cut is a scroll with no graph—not even a line of identity—on its first (outer) area. So the rule is restated as follows: R5. *The double cut may be inserted around or removed (where it occurs) from any graph on any area. And these transformations will not be prevented by the presence of ligatures passing from outside the outer cut to inside the inner cut.*

It must be emphasized that the new application of R5 is restricted to cases in which lines of identity only are on the outer area of the double cut, and where these lines do not terminate on that area but pass all the way from outside the outer cut to inside the inner cut. The point is that there are only two ways in which a cut may affect the significance of a line of identity in Beta: (1) the interpretation of a line as denoting 'something' or 'everything' is determined by the cut which encloses the outermost extremity of that line; and (2) the significance of a line is altered from identity to non-identity when it passes entirely through the area of a cut. It is clear that the addition or removal of the double cut in accordance with the restriction just stated cannot possibly change anything in regard to (1) or (2).

As I mentioned before, the EG conventions and rules are collected together in Appendix 3. Now we proceed to some illustrations of Beta.

4.3 FURTHER ILLUSTRATIONS

(1) In section 3 of Chapter 2 a graph of the proposition 'Every mother loves some child of hers' was presented. That one was drawn by Peirce in 1893 in a chapter of his *Grand Logic*, and modeled (as he states) after Kempe's system of diagrams. I promised at that time to present an existential graph of the same proposition after Beta had been introduced, and here is the graph:

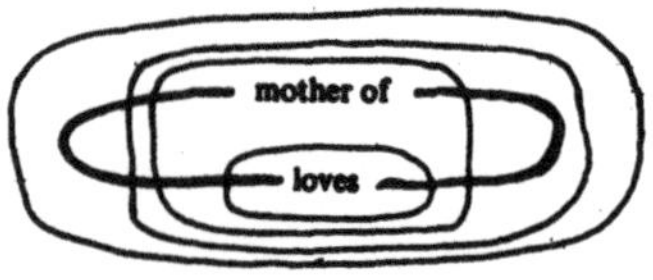

This is obviously far less complicated than the graph of 1893. Let *i* be the individual denoted by the line of identity at the left (the mother), and *j* the individual denoted by the other line (the child), and read the graph as follows: 'Take any individual you please, say *i*, there is an individual *j*, such that, if *i* is mother of *j*, then *i* loves *j*'.

(2) The syllogism Barbara was something of a favorite of Peirce's, and the graphical analysis of it appears in several places throughout his papers (see, for instance, 4.571, 5.147, Ms 500, Ms AM 806.5*, and Ms L 231, pp. 21-22). This is the AAA syllogism in the first figure: 'All F is G, and all G is H; therefore, all F is H'. In terms of EG, we must justify the transformation of the graph

into the graph

1. 

Premisses.

2. 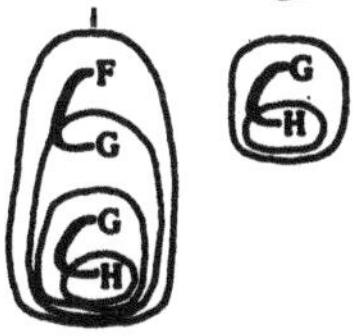

1 by R3. Iteration within two cuts of the premiss 'All G is H' which, in step 1, occurs on SA.

3. 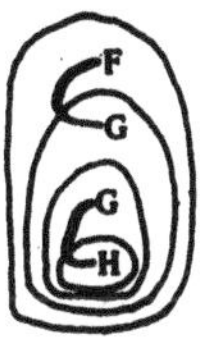

2 by R1. Erasure of the occurrence of 'All G is H' which, in step 2, occurs on SA and thus—by convention—is evenly enclosed.

4. 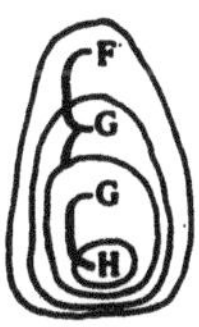

3 by R3 (a).

5. 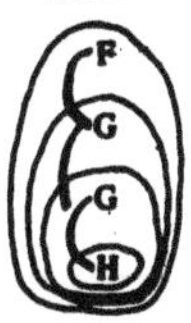

4 by R3 (b).

6. 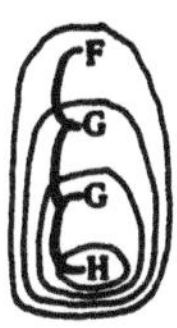

5 by R2.

7. 

6 by R4. Deiteration of the thrice enclosed — G since another occurrence of — G is only twice enclosed.

8. 

7 by R5.

9. 

8 by R1.

(3) Next we give Beta proofs of the two axioms of identity: (x)[x = x], 'Everything is identical to itself'; and (x) (y)[x = y ⊃. Fx ⊃ Fy], 'Take any two objects x and y, if they are identical then whatever is true of x is true of, y' (Quine [1959], 213; Church [1956 a], 281).

A graph of the proposition 'Everything is identical to itself' is 
Here is the proof:

1.  R5.

2.  1 by R2.

3.  2 by R3(b).

4.  3 by R3(d).

A graph of the second axiom of identity can be built up by first diagramming (x) (y) [Fx ⊃ Fy], 'For every x and y, what is true of x is true of y'. The top line is the x-line:

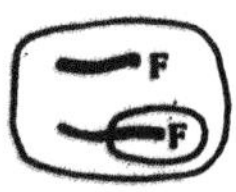

Since both spots are to appear in the consequent of the conditional proposition we are constructing, we add two cuts, being careful to keep the lines of identity precisely once-enclosed:

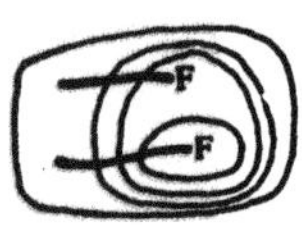

Now by joining the lines in the first area of the nest we add the condition 'x is identical to y':

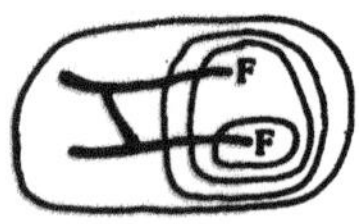

Here is the proof:

1.  R5.

2. 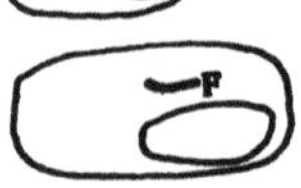

1 by R2.

3. 

2 by R3.

4. 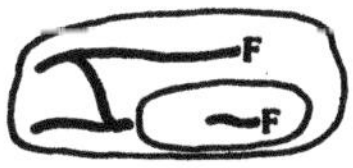

3 by R3(a).

5. 

4 by R3(b).

6. 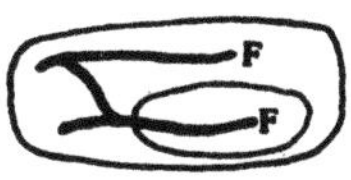

5 by R3(c).

Now add double cuts by R5.

5

GAMMA

The Gamma part of EG corresponds, roughly, to second (and higher) order functional calculi, and to modal logic. Because it was never completed, it is occasionally difficult to be sure of just what Peirce was up to. Nevertheless, the attempt to understand Gamma is exciting and valuable, partly because the development of this part of the system parallels late developments in other parts of Peirce's philosophy. A full account of these parallels (as well as those to be mentioned in Chapter 6) must wait for later and separate treatment.

By means of this new section of EG Peirce wanted to take account of abstractions, including qualities and relations and graphs themselves as subjects to be reasoned about. To do this he invented new spots, new ligatures, new cuts, and new sheets of assertion. It is one of the Gamma cuts that provides a first graphical treatment of modality.

The fullest exposition of Gamma was prepared for the Lowell Lectures, delivered in November and December of 1903. The surviving manuscripts, Mss 447-478, contain also a full and careful exposition of the rest of EG (which was heavily used in the two preceding chapters). This account of the graphs was based in large part upon a manuscript entitled '*Logical Tracts. No. 2*', written perhaps the same year.[1] These accounts are the major sources for the present chapter.

5.1 THE 1903 ACCOUNT OF ABSTRACTIONS

On August 4, 1898, Peirce wrote this heading on page 128r of his *Logic Notebook*: "We now come to An Extension of Existential Graphs, permitting Abstraction". Abstraction, he says, "consists in asserting that a given sign is applicable instead of merely applying it". Instead of merely saying

1 Ms 492, published in large part in 4.418-509. The dependence claim is partly based on the last page of Ms 450, one of the Lowell Lectures, which refers to page numbers of Ms 492.

'Napoleon was a great man' we say 'Napoleon was a man to whom the term "great" could be applied'. Such assertions are easy to come by in ordinary language, whose abstractive facility renders it "more logically powerful than any algebra of logic hitherto developed" (*LN*, p. 127v). Entries in this manuscript show that Peirce examined the topic for five more days. He introduced two special symbols, one to signify ordered pairs and the other to signify either membership in a collection or possession of a character. Thus, the graph of Fig. 1 means 'A is the first, B the second of the ordered pair C'. Fig. 2 means 'A belongs to the general unordered collection B' or 'A possesses the character B'. Fig. 3, combining the two symbols, reads 'A stands to B in the relation C'.[2]

Fig. 1

Fig. 2

Fig. 3

By means of these symbols Peirce was able to express such propositions as 'Any two things (or the same thing) form a pair' (Fig. 4), 'No two things form more than one pair' (Fig. 5), and 'Given any two things, there is some character which one possesses and the other does not' (Fig. 6) (*LN*, p. 131r). At this early date, Peirce was content to use the standard line of identity (even for Fig. 5 where it represents two types or categories of things), and special spots are given in English, as 'is a sequence'. He recognized that the universe of discourse required attention, but "for simplicity in these graphs different universes are not distinguished" (*Ibid.*). I wonder how he would have distinguished them had he decided to do so.

Fig. 4

Fig. 5

Fig. 6

The kind of abstraction defined above is what Peirce called 'hypostatic' abstraction.[3] By 1903 he regarded it as an inference, necessary and immediate, whose conclusion refers to something which is not referred to by the premiss. This new something is itself called an abstraction, or an *ens rationis*, a "creation of the mind" (Ms 458, p. 14); and its being consists in the truth of what the premiss asserts about something else.[4] In the Napo-

2 The interpretation of these symbols (and of many others in the pages to come) suppresses the phrase 'there exists' for each unenclosed line of identity. By now the reader must be quite familiar with the line as quantifier.

3 4.549, 346. It is to be distinguished from 'precisive' abstraction, which is the mental separation of items resulting from concentrating on one element and neglecting others. 1.549, 4.463.

4 4.463, Ms 460, p. 20 verso. Cf. Ms 462, pp. 34-36; 4.235.

leon example, the term 'great' is not referred to in the premiss, although it is used there; it is referred to in the conclusion. That conclusion might have been stated differently to make the distinction even clearer: 'Napoleon possessed greatness'. Here the new something, the resultant abstraction, is greatness. It is not referred to in the premiss, but its being consists in the truth of what the premiss asserts about Napoleon.

Sometimes, little is gained by the more complicated expression which results from hypostatic abstraction. This is the case with one of Peirce's favorite examples (Ms 467, p. 66; 4.463):

Opium puts people to sleep.
Hence, opium has dormitive virtue.

Yet, according to Peirce, the gist of mathematical reasoning lies in so "turning what one may call adjective elements of thought into substantive objects of thought" (Ms 462, p. 48). Consider, for instance, the following "logical proportions" (Ms 467, p. 78):

A particle moves A particle describes a line	A pear is ripe A pear possesses ripeness

A filament moves from its place A filament generates a surface	Ripeness is more or less Ripeness possesses degree

In each column of each set of boxes, the forms of statement in the upper box have the same relation to the forms of statement under them. And that relation is the relation of premiss to conclusion in an inference by hypostatic abstraction. Instead of saying 'A particle moves' we introduce an abstraction and say 'A particle describes a line'. In the same way we introduce the notions of dormitive virtue and ripeness. We can go further, imagining a filament to occupy the whole line at once, and then to move all at once, from which we get the idea of the filament generating a surface. Nor is it necessary to stop here.

We may imagine a material *film* to occupy the whole surface at once, and may imagine that the *film* moves in such a way that at each moment it is quitting the surface at which it is just arriving; and if it is not restricted to the space of our ordinary intuition we may say that the character of its motion determines a particular space, tridimensional at each point of it. This space, probably much more peculiar than the simple space we know, might, for aught we can see, be occupied all at once by a *body*. And we cannot see why this body should not move so that at every instant it should altogether quit the space that it at that instant occupies; and so on [Ms 467, pp. 68-69].

This fertile type of reasoning has given us the mathematical concepts of number, collection, series, relation; it teaches us to treat operations themselves as things to be added, multiplied, and raised to powers (Ms 462, p. 48). It has given us the philosophical concepts of truth, humanity, justice, and so on (see Ms 469, p. 10). Small wonder that Peirce wanted his graphs to be able to deal with such reasoning.

While postponing any general treatment of the different universes of discourse which abstractions involve, we can do quite a bit with Beta graphs. Fig. 7, for example, is a graphical definition of necessary reasoning as "that whose conclusion is true of whatever state of things there may be in which the premiss is true" (Ms 459, p. 24).

Fig. 7

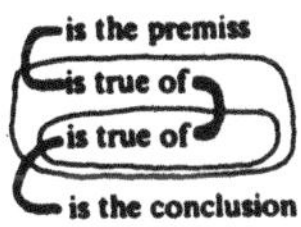

The pure mathematician generalizes this proposition, substituting for the logical terms "the indefinite symbols *x*, *y*, *z*, which are to mean whatever they may mean; and he thus gets this graph, which is precisely the graph of inclusion"(Ms 459, p. 25).

Fig. 8

Again, in setting up a proof that there is no multitude intermediate between 2 and 3, Peirce scribed the graphs of Nullity, Unity, Twoness, and Threeness. He used only cuts and the standard line of identity, but there is this special proviso, that in each graph the universe of discourse is to be the members of the collection there dealt with. The derivation is rather interesting.

Nullity will be expressed by  which is of course the pseudograph. For a universe with nothing in it is absurd. I add one to this. That is I express that there is something such that if anything is different from this the graph of nullity holds for it. That is  or . It is the graph of *Unity*. I add a unit to this. That is I express that there is something other than what the graph of unity asserts, but for all that is not this that graph holds. This gives . It is the graph of Twoness. I again add a unit, asserting that there is something else but that apart from this new unit the graph of Twoness holds. This gives

which is the graph of Threeness [Ms 469, p. 68].

Nevertheless, if the graphs are to fulfill their intended purpose and provide a diagram of reasoning, and an iconic diagram at that (see Chapter 7, section 24), a special symbolism must be provided for the reasoning about abstractions. Peirce went about this with enthusiasm, producing for Gamma a "great wealth of new signs" (4.512). As he pointed out and as we shall see, however, none of these symbols is of an essentially different kind from those of Alpha and Beta. Rather we find new versions each of spots, lines, cuts, and sheets of assertion.

5.11 *The Potentials*

In *Logical Tracts. No. 2* Peirce provided two conventions for dealing with abstractions. One of them, No. 14, will be dealt with in 5.13 below. The other, No. 13 (4.470; cf. Ms S 28, p. 51), introduced the letters $\varrho_0, \varrho_1, \varrho_2, \varrho_3$, and so on, as spots to be read as follows:

A is a proposition or fact.

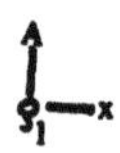

x has the monadic quality A.

x is in the dyadic relation A to y.

x is in the triadic relation A to y for z.

Rhos of higher adicity can be scribed in similar fashion, reserving the vertical line for the abstraction, and reading the other lines clockwise when they are on the same area (read them endoporeutically if they are not). Thus, Fig. 1 expresses the proposition 'There is a relation (unspecified) that every man bears to some woman'.

Fig. 1

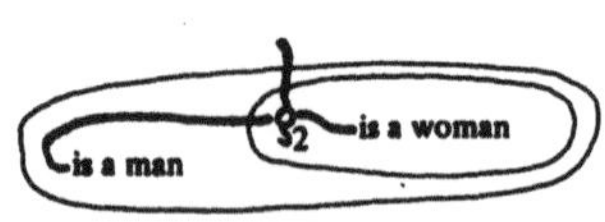

While preparing his Lowell Lectures in 1903, Peirce improved this convention by replacing the rhos with a new and larger set of spot-symbols, which he called 'the Potentials'. This seemed a good name for symbols used to signify logical possibilities (Ms 478, p. 159). Here they are, as given in Ms 467, p. 80.[5]

A—p	A is a primary individual.
A—q	A is a monadic character, or quality.
A—r	A is a dyadic relation.
A—s	A is a legisign, or triadic relation.
	A is a graph, or proposition, or fact.
	B possesses the quality A.
	B is in the dyadic relation A to C.
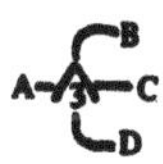	B is in the triadic relation A to C for D.

The selectives (capital letters) are used to facilitate the reading of the graphs, and also to call attention to the fact that the lines of identity to the left of the potentials are "peculiar" in that they denote abstractions, not existing individuals as in Beta.

One of the ways Peirce used the potentials was to express propositions having to do with sequences. Let us watch him build up a graph expressing the relation of coming after, that is, the relation of posteriority. He begins with Fig. 2, which expresses the relation of being at least as late as.

Fig. 2

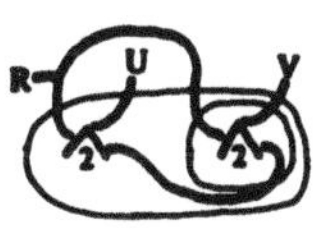

We can see that this does the job if we choose R to be the relation 'less than', and let the other lines denote numbers. For then the graph asserts that V is less than each number that U is less than. So that in counting, U is at least as late as V. It is not excluded by this graph that U and V are the same number.

Peirce says that he usually calls the relation of Fig. 2 "inclusion of correlates" because the graph implies that "everything that U stands in any fixed relation to is included among the things to which V stands in that same relation" (Ms 459, p. 20).

5 Published in 4.524 (cf. 4.409). Incidentally, portions of Fig. 194 in 4.526 are difficult to make out. You may wish to note the following in your copy of CP: in the first area Peirce scribed Qq; in the second, Rr; in the third, two instances of Q↑; the other eight spots are instances of R$\underset{2}{\wedge}$, four of which are enclosed in cuts containing nothing else (aside from the lines of identity).

Peirce now combines the graph of Fig. 2 with its negative to produce a graph of the relation of coming after (Fig. 3).

Fig. 3

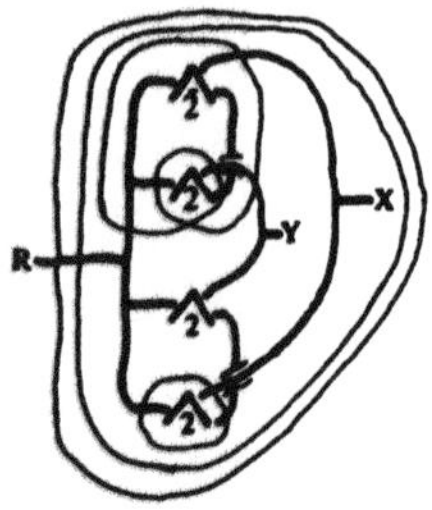

That is, "to say that X is posterior to Y is to say neither more nor less than that there is a relation R in which Y stands to whatever X stands in that relation to as well as to something to which X does not stand in that relation" (Ms 459, p. 19). R can be any dyadic relation, but it must be fixed in advance–which is indicated by the double cut. The reason for this requirement is that changing from one relation to another would generally be to change the order of sequence. The reader can convince himself of this by reading R as the relation 'greater than'.

Peirce was working up a rule for potentials as he prepared his Lowell Lectures; he gave it two clauses in one place (Ms 459, pp. 30-31), four clauses in another, and then stated that the rule had not yet been "fully formulated" (Ms 478, pp. 159-161). The complete lack of illustrations makes it difficult to discover just what he had in mind for this rule, other than embedding in EG his developing doctrine of substantive possibility. Qualities and relations, which are not existent objects, but are rather general respects in which existent objects *might* agree or differ, are examples of substantive possibilities. Such possibilities are 'prior to existence' in the sense that non-existence does not necessarily prove non-possibility, but non-possibility does prove non-existence (Ms 459, p. 29). The rule which purportedly follows from these truths contains the following clause:

> Any graph which does not relate to what exists but only to pure substantive possibilities of the same order is true if the outermost parts of its innermost ligature is enclosed in an even number of cuts, but is false if that number be odd [Ms 459, p. 30].

We note first that this clause is restricted to the potentials q, r, s, and Λ_0. (The potential p is ruled out because it signifies a primary individual, not an abstraction; and Λ_1, Λ_2, and Λ_3 are excluded because they involve ligatures of existent individuals.) We note next that the effect of it is to allow the assertion of such propositions as — r and —Λ_0, 'there is a dyadic relation' and 'there is a proposition'. And it rejects such propositions as (—q) and

, 'nothing is a quality' and 'everything is a proposition'. I have not found an explanation of what is meant by 'substantive possibilities of the same order', but perhaps it is simply a kind of ranking used to prevent category mistakes.

5.12 *Graphs of Graphs*

A second category of Gamma graphs contains those which enable us to reason about graphs themselves. The various symbols for this metalanguage are presented in two groups: those for Alpha graphs and those for Beta graphs. They are taken from Ms 467 (the source of 4.528-529) and Ms 468 (the source of 4.529n.*).

Gamma Expressions of Alpha Graphs

X—A	X is the sheet of assertion (SA).
X—α	X is an area.
X—γ	X is a graph, or graph-instance.
X—τ	X is a point.
X—φ	X is a permission.
X—χ	X is a fact.
X—μ	X is a blank.
X—κ	X is an enclosure.
Y—λ—X	X is placed on Y.
X—(Y)	X precisely expresses Y.[6]
X—ι—Y	X is the area of the enclosure Y.
X—θ—Y	X is a point of Y.
X—δ—Y	The graph-instance denoted by X contains as a part of it the instance Y.
X—ζ—Y	X is an instance of the same graph of which Y is an instance, or X is equivalent to Y.
X—σ<Y, Z	X carries Y as its entire graph in so far as it is of the nature of Z to make it do so.

6 It is necessary to place Y in the saw rim "because in thus speaking of a sign *materialiter*, as they said in the middle ages, we require that it should have a hook that it has not got" 4.528. This saw rim is different from the one to be introduced in 5.13 below.

In these manuscripts Peirce used the term 'graph-replica'. I have changed it to 'graph-instance' in accordance with his remark of 4.395n.1, where he concedes priority to Kempe regarding the term 'replica'.

Our first illustrations express facts which relate to the physical possibilities and necessities concerning graphs. Peirce distinguished thirty of these in Ms 511, but the following graphs are from Ms 468.

The sheet is an area.

There is but one SA. (Take any two objects: if each is SA, then they are identical.)

Every blank is a graph-instance.

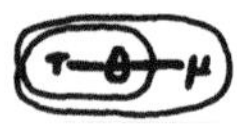

There is a point on any blank.

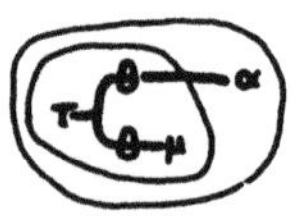

On every area there is a blank. (Take anything you please, and if it is an area, then there is a point and a blank such that that point is both a point of that blank and a point of the area.)

Every enclosure is a graph-instance.

Every enclosure has an area.

No enclosure has more than one area.

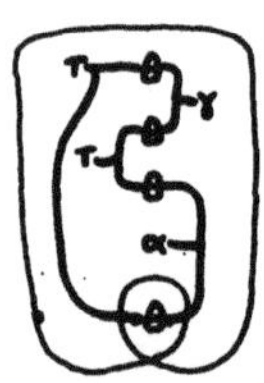

A graph is wholly in one area. (Take any graph, any two points of that graph, and any area; if one of the points is on that area, then so is the other.)

Here is an approach to the rule of the double cut:

It is permitted to place on SA as the entire graph an enclosure on whose area an enclosure is placed as a fact.

Fig. 1 is an Alpha graph which expresses the proposition 'If it hails then it is cold'. A graph asserting that Fig. 1 is scribed on SA is given in Fig. 2:

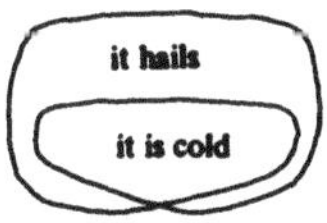

Fig. 1

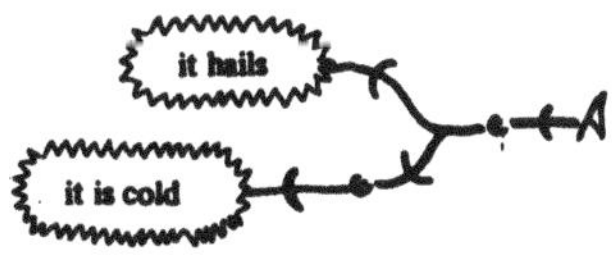

Fig. 2

Let us dissect this graph to ensure that it is correct.[7] Fig. 3 asserts that a graph expressing 'It hails' is scribed on X:

Fig. 3

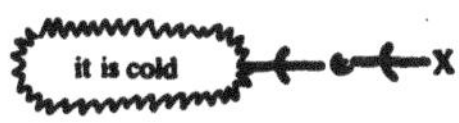

Fig. 4

Fig. 5

Fig. 4 asserts that a graph expressing 'It is cold' is scribed on an area of an enclosure which is scribed on X; and Fig. 5 asserts that X is the area of an enclosure which is scribed on SA. Joining the lines of each which denote the area X, gives the graph of Fig. 2.

If it is required to say that the graph of Fig. 1 and nothing else is scribed on SA, we introduce two new spots: Fig. 5, which means 'X and nothing

Fig. 5

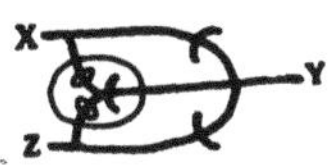

Fig. 6

else is scribed on Y' (literally, 'X is scribed on Y and it is false that there is something other than X which is scribed on Y'), and Fig. 6, which means 'X and Z and nothing else are scribed on Y'. The required graph is obtained by placing these spots into Fig. 2 in such a way as to produce Fig. 7:

Fig. 7

7 Neither Fig. 196 of 4.528 nor the original graph in Ms 467 (p. 92) is quite correct.

Gamma Expressions of Beta Graphs

X is a dot.

X is a line of identity.

X is a point of teridentity.

Y is a ligature whose outermost part is on X.

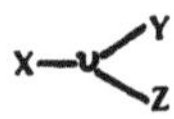

X is a line of identity having its terminals at Y and Z.

g is expressed by a monad spot on X whose hook is joined to the ligature Y on X.

g is expressed by a dyad graph on X whose first and second hooks respectively are joined on X to the ligatures Y and Z.

Graphs similar to the last two can be constructed for other polyads. The rule for reading the individuals denoted by the lines of identity other than the one positioned at 9 o'clock, is to take them in their order clockwise.

The following illustrations express additional facts about EG.

Every line of identity is a graph.

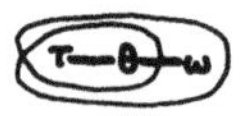

There is a point in a dot.

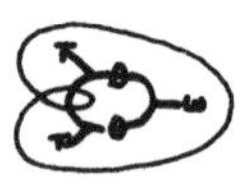

If in any dot there is a point A or a point B, these are identical.

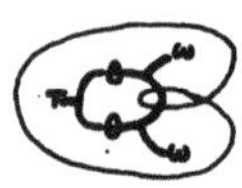

No point is in two dots. (Take any point and any two dots; if the point is on both dots, then the two dots are one and the same.)

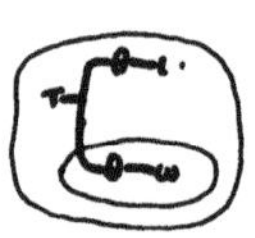

At whatever point on a line of identity there is a dot. (Take any line of identity and any point, call it A; if A is a point of the line, then there is a dot such that A is a point of that dot.)

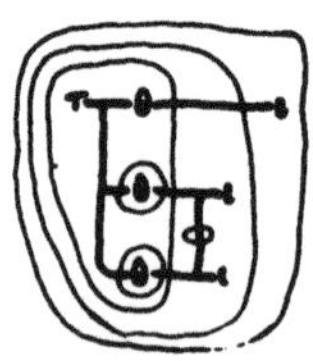

If there is a line of identity there are two individual lines of identity such that every point of the [original] line is either in the one or in the other.

This last graph asserts that a line of identity may be regarded as consisting in two lines of identity. This is presupposed in the rule of erasure which says that any portion of an unenclosed line may be erased; for if you erase a portion in the middle, you obtain two lines.

A few more graphs of graphs will be presented in section 5.2, after the broken cut has been introduced.

5.13 *The Three Gamma Rims*

As remarked earlier, the lines of identity to the left of the potentials are peculiar in that they denote abstractions and not existent individuals; the lines in the graphs of graphs are also peculiar, signifying graph-instances. The potentials and the special Gamma spots do call this to the reader's attention, but Peirce invested a lot of ingenuity in devising ways to make it even more noticeable. We consider first Convention No. 14 of *Logical Tracts. No. 2.*

> The line of identity representing an *ens rationis* may be placed between two rows of dots, or it may be drawn in ink of another colour, and any graph, which is to be spoken of as a thing, may be enclosed in a dotted oval with a dotted line attached to it. Other *entia rationis* may be treated in the same way, the patterns of the dotting being varied for those of different category [4.471].

Early drafts of the manuscript show that Peirce at first hesitated to include the conventions on abstractions (and those on selectives) because they had not been "thoroughly studied", and because he was not yet sure that they should belong to the system at all (Ms 492, p. 63). His first attempts at abstraction symbols were simple dotted boundaries (*Ibid.*, p. 79), and he introduced a rule of inference which permits us to transform any graph, say A (Fig. 1), into another graph which asserts that A is true (Fig. 2); the rule permits the reverse transformation as well.

A

Fig. 1

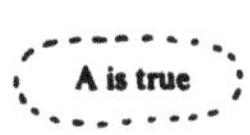

Fig. 2

The diagraming itself "may be done in different ways", which indicates that the reasoner is expected to invent symbols on his own according to his need.[8] The rule is stated generally without reference to any particular symbolism: "[An Alpha or Beta] graph may be replaced by an equivalent graph introducing an *ens rationis*, and any graph involving an *ens rationis* may be replaced by an equivalent [Alpha or Beta] graph" (Ms 492, p. 112).

8 Ms 492, p. 111. From Ms 693, p. 282: "*The Gamma Part* supposes the reasoner to invent for himself such additional kinds of signs as he may find desirable". By 'additional' is meant additional to the Alpha and Beta signs, which indicates that Ms 693 was written earlier than Ms 492, before special Gamma signs began to appear.

Figs. 1 and 2 are not "strictly speaking" equivalent, however; and the recognition of this (*Ibid.*, p. 114) may have urged caution upon Peirce. For the final version of the manuscript contains the rule in weakened form (4.507, third clause of Note A), and only one example each of the dotted line and the dotted cut.

First, the illustration of the dotted line:

Fig. 3

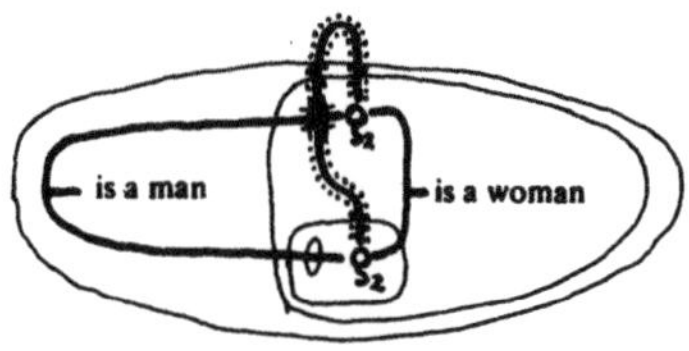

The graph asserts that there is a relation in which every man stands to some woman, a relation in which no other man stands to the same woman. This means that there is a woman corresponding to every man, or, "there are at least as many women as men".[9]

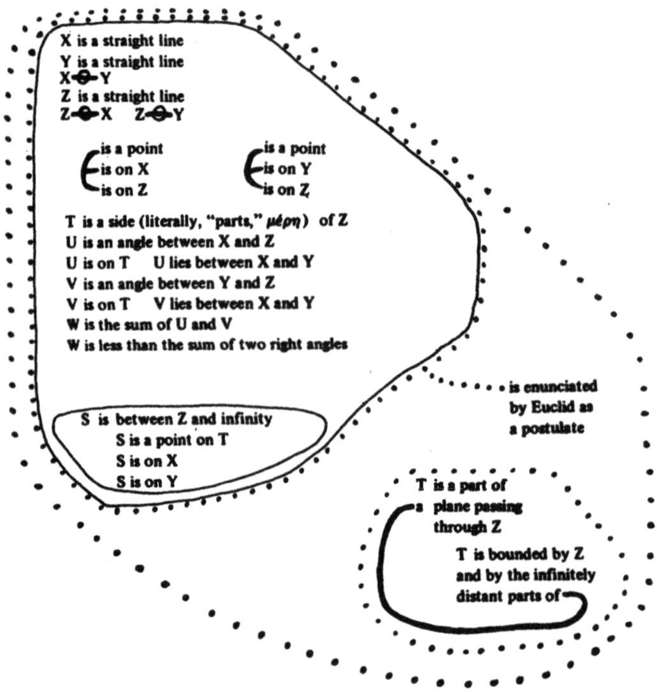

Fig. 4

9 The reproduction of this graph in 4.470 lacks the bridge which is necessary to keep the two lines distinct. Also, the tail on 'is a man' should extend to the oddly enclosed line as shown here. Ms 492, p. 82.

As an example of the use of the dotted cut or oval, Peirce presented a graphical analysis of a postulate given by Euclid: "If a straight line incident upon two straight lines makes the angles that are inside and on the same parts less than two right angles, then those two straight lines being prolonged to infinity shall meet on what parts the angles were less than two right angles" (Ms 492, p. 84). The graph as Peirce drew it is given in Fig. 4 (correcting two small errors of Fig. 120 in 4.471). The dotted oval which encloses the entire graph signifies that the graph is about abstractions; and the two once-enclosed dotted ovals signify the same thing for the parts of the graph. One of these parts, attached to the dotted line, is the postulate asserted by Euclid; the other part states the conditions under which the postulate holds. The graph can be read, with some attention to detail, as follows:

Euclid asserts as a postulate that:

If X, Y, and Z are distinct straight lines; and if there is a point common to X and Z, and a point common to Y and Z [That is to say, if X and Z intersect and if Y and Z intersect]; and if T is a side [We are to imagine that the line Z lies on a plane, cutting that plane into two parts or sides; T is one of those sides.] of Z, U an angle made by X and Z lying on T between the lines X and Y, and V an angle made by Y and Z lying on T between the lines X and Y; and if W, the sum of the angles U and V, is less than the sum of two right angles—the situation as described so far is shown in Fig. 5—

Fig. 5

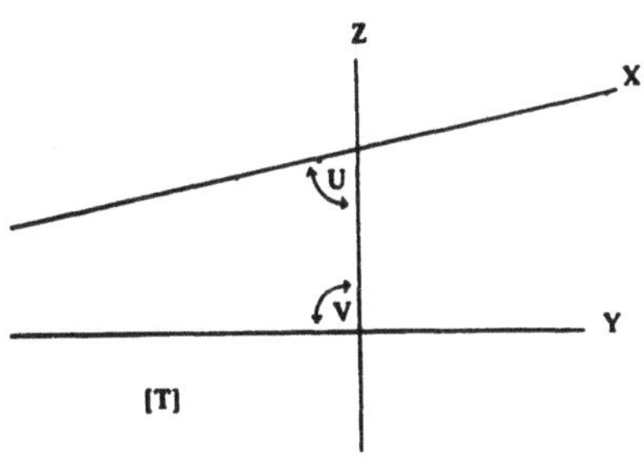

Then: S is a point on T common to lines X and Y, and S occurs on T somewhere between the line Z and the infinitely distant parts of T (where T is a part of a plane passing through the line Z, that part namely that is bounded by Z and by the infinitely distant parts of the plane of which it is a part).

As he worked on his Lowell Lectures, Peirce tried to expand the graphical treatment of abstractions. He spoke of using the colored lines of Convention 14 along with similarly colored selective capital letters (a letter for each different character). But (felt-tipped pens not being then available) he judged that this "would involve too much delay and trouble" for ordinary work (Ms 464(s), p. 80). For a time he was to shift his attention to specially designed cuts and spots, but color returned with a vengeance in 1906 (Chapter 6 below).

At first he worked with the single dotted cut of *Logical Tracts. No. 2.*[10] This cut is used with potentials in the third Lowell Lecture "to denote the single character which consists in the logical possibility of the rhema written within it" (Ms 464(s), p. 84). Fig. 6 gives an example which contains an interpreting Beta graph. The Gamma part of Fig. 6 asserts that some man

Fig. 6

possesses the character of being left-handed. This is another way of saying what the Beta part says, and although the graphs are not 'strictly speaking' equivalent, it should be possible to infer each from the other. The rule of inference quoted at the bottom of p. 75 permits this.

At some stage of the work, the unenclosed potential in Fig. 6 was replaced by a second dotted cut, the 'peculair line' (to the left of the potentials) was changed to a dotted line, and the inner potential disappeared. Thus we

Fig. 7

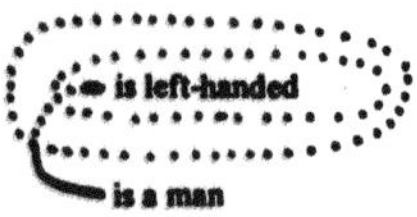

have a new graph, given in Fig. 7, of the proposition expressed in Fig. 6. In addition, the name 'cut' was replaced by the name 'rim', and a rim was defined as "an oval line making it, with its contents, the expression either of a rheme or a proper name of an *ens rationis*" (4.411). In the final stage of this development, the system contained three rims with associated tags or lines: a double dotted cut with a dotted line, a wavy cut with a heavy wavy line, and a saw cut (different from that of section 5.12) with a saw line. The function of these lines was to identify the subjects of the abstractions. A drawing of the saw cut occurs on a rejected page of the *Syllabus* (Ms 478, p. 148), and what is called a wavy cut is used for emphasis (with no Gamma

The individual X has the character of being a B.

A is one of the collections of X's.

A is the collection of all X's.

10 See for instance early drafts of Ms 478, p. 148.

function) in a letter to William James written in August, 1905 (Ms L 224). Above are the three rims, as well as I can make them out.

The saw rim functions in much the same way as the capped variable prefix does in *Principia Mathematica*. If Fx is 'x is a man', $\hat{x}(Fx)$ will be the class of all men.[11] In the graphs, nothing precisely like Fx can be expressed, since free individual variables do not occur in EG. But from 'Something is a man' in Fig. 8 we obtain a graph which denotes the collection of all men by applying the saw rim and attaching the saw line to the heavy dot, as in Fig. 9. Fig. 10 employs the wavy rim and line to signify an unspecified collection of men, not necessarily the unique collection of all men.

Fig. 8

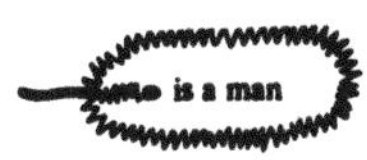

Fig. 9

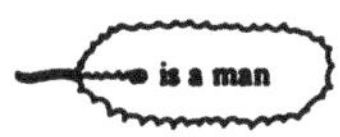

Fig. 10

The wavy rim and saw rim are used together in Fig. 11.

Fig. 11

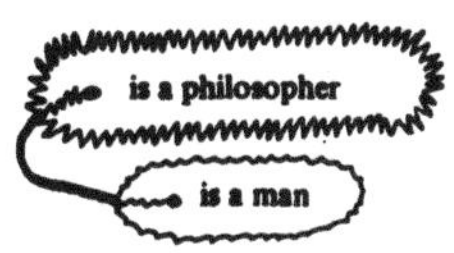

The proposition is that the collection of philosophers is one of the collections of men.

The wavy and saw rims may contain more than one wavy or saw line, as in Figs. 12 and 13, where X, Y, and Z signify objects or sets of objects.

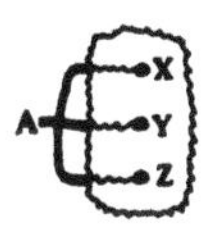

Fig. 12

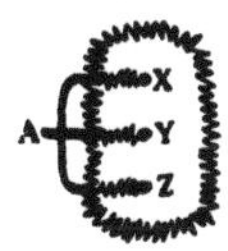

Fig. 13

The difference between these graphs is that the collection of Fig. 13 is the unique collection of all the X's, Y's, and Z's in existence; that of Fig. 12 is one of the possible collections of X's, Y's, and Z's.[12]

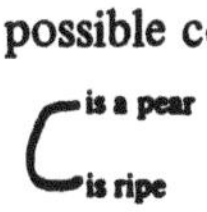

Fig. 14

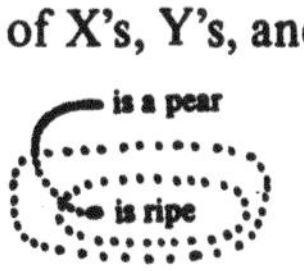

Fig. 15

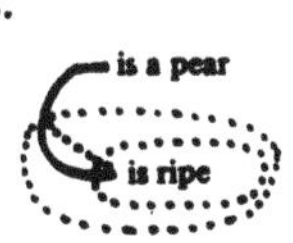

Fig. 16

11 In *Principia*, the 'all' is given in the definition of a class as 'all the objects satisfying some propositional function' (Whitehead and Russell, 1927, volume I, p. 23).

12 For example, if there are exactly two X's, two Y's, and two Z's, then there are 27 possible collections containing at least one each of X, Y, and Z, and the wavy rim can be used to denote any one of these. One of the 27 is the collection of both X's, both Y's, and both Z's; and this is the one denoted by the saw rim.

Consider the dotted rim once more. Fig. 14 asserts 'A pear is ripe', and Fig. 15 asserts that a pear possesses ripeness. If we ignore the extra length of line in Fig. 16, and take this graph to be equivalent to that of Fig. 15—with regard to the significance of the dotted rim—then we have a way to diagram inferences by hypostatic abstraction that involves no change of the original graph, but simply an addition to it. We can imagine that applying the dotted rim to a part of a graph is something like placing a grid or a stencil over a coded message; in each case something is brought to our attention that might otherwise have been overlooked.

Fig. 17 is the proposition 'Opium puts people to sleep'.

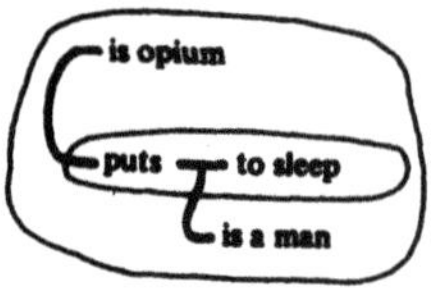

Fig. 17

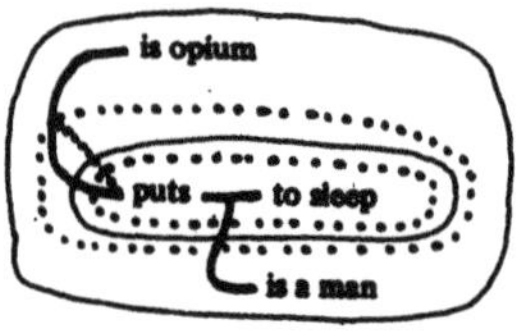

Fig. 18

Inserting the dotted rim produces Fig. 18 which means 'Opium has dormitive virtue'.

A dotted cut may contain more than one dotted line and thus express more than one character. By adding a dotted line to the graph of Fig. 18 we add to that graph the proposition 'Man is susceptible to soporifics'.

Fig. 19

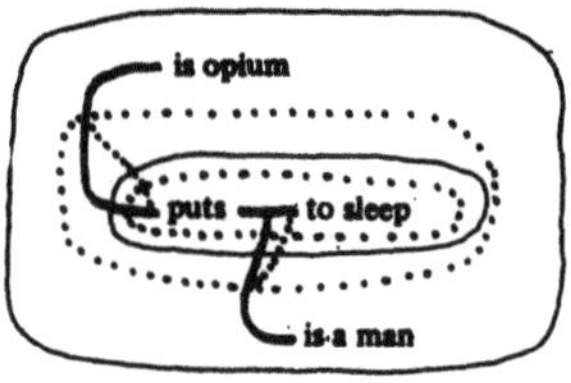

5.2 THE 1903 ACCOUNT OF MODALITY

We have seen that as early as 1898 Peirce thought of making distinctions between different universes of discourse involved in his graphical assertions. He postponed any symbolic treatment of this issue, simply stating the requisite conditions whenever he thought it necessary. The new symbols of 1903, the potentials, the graphs of graphs, the Gamma rims and the dotted line, were all attempts to classify different objects involved in reasoning. He did not give up on the idea of treating the universe directly by means of improvements on the sheet of assertion, and by 1903 he had some ideas to report. They are found in the fourth of his Lowell Lectures.

The essential improvement is to replace the standard SA by "a book of separate sheets, tacked together at points, if not otherwise connected" (4.512). The upper sheet will be the standard SA and will represent, as always, "a universe of existent individuals", different parts of the surface representing facts about, or true propositions asserted of, that universe. By means of the cuts–you may imagine that we are actually to cut out a piece of the upper surface–we get down to the other sheets in the book, and these other sheets "represent altogether different universes with which our discourse has to do" (4.514). On these new areas may be placed "conceived propositions which are not realized" (4.512) and reasonings involving qualities (4.514). Now qualities are logical possibilities, and "possibilities are general, and no multitude can exhaust the narrowest kind of a general" (*Ibid.*). Nevertheless, an approach to a representation of this "entire universe of logical possibilities" is gained by this addition of depth to the sheet of assertion.

These are intriguing ideas. Peirce admitted that the lecture would not afford time enough for their full development; more than that, he indicated that his own practice was restricted to a small number of ideas which he found "convenient to work with", and the book of sheets was not yet one of these (4.514). He finally decided to omit this material (from the ellipsis marked in 4.512 to the end of 4.514) and substitute instead this single page:

If I were to expound to you fully the theoretically needed new forms of spots, cuts, and ligatures that are required in the gamma part, you would find the complexity of it, – presented in the hurried way that would be necessary, – to be not only tedious but also confusing.

It will be better to give you examples of what I have found most useful, and leave it to you to study out the rest if you care to do so. I shall have a printed syllabus ready for distribution at the next lecture which will be a great help; but even in that I cannot go into the long explanations that would be needed to expound the theory of the gamma part [Ms 467, p. 21].

The insert makes no mention of the book of separate sheets; and no mention of it is made in the *Syllabus*, either. But as we shall see in the next chapter, this idea is central to Peirce's last revision of EG.

The fourth lecture then continues with an exposition of the broken cut, published in 4.515ff. Whereas the Gamma symbols already introduced enable us to make abstractions or logical possibilities the subjects of discourse (Ms 460, p. 21), the broken cut will enable us to 'predicate' possibility–and the other modal notions–of propositions and facts. Modality, said Peirce in about 1902, "is the logical qualification of a proposition or its copula, or the corresponding qualification of a fact or its form, in the ways expressed by the modes *possibile, impossibile, contingens, necessarium*"

(2.382; cf. 2.323). There are then four modes to be expressed in terms of the new symbol.[13]

It is interesting that in an early draft of the *Syllabus*, the broken cut was listed with the potentials, and these illustrations were given (Ms 478, p. 147):

is necessary

possible that necessary

possible

The final draft contains no illustration of the cut but simply describes it as having "many little interruptions aggregating about half its length", and then presents the following, which is the tenth convention in our list: C10. *The broken cut expresses that the entire graph on its area is logically contingent (non-necessary)* (4.410). Thus, Fig. 1 does not assert that it does not rain, but that it is not necessary that it rains, or 'It is possible that it does not rain'.

Fig. 1

The *Syllabus* contains the most explicit statement of rules for the broken cut. There we learn that Peirce did not allow a graph to be iterated or deiterated across a broken cut, so that R3 and R4 do not hold for this cut. R5, the rule of the double cut, does not hold either, although a double broken cut may be scribed unenclosed as long as *both* of its areas are blank ("yet, owing to the failure of Iteration and Deiteration, this leads to nothing") (Ms 478, p. 158).

There is one new rule:

R6. The rule of cut conversion. (a) *An evenly enclosed standard cut may be transformed (by being half erased) into a broken cut*; *and* (b) *an oddly enclosed broken cut may be transformed (by being filled up) into a standard cut.*

Fig. 2

Fig. 3

g

Fig. 4

[13] For the record, the four modes just listed are expressed in the upcoming Figs. 5, 6, 1, and 2, respectively.

The above mentioned fourth Lowell Lecture, the only exposition of the broken cut that I have found, contains several new inferences. First, from 'g is necessarily true' (Fig. 2) we can derive the proposition 'g is true' (Fig. 4). That is, from Fig. 2 (literally, 'It is false that g is possibly false') we obtain Fig. 3 by R6(b); and Fig. 4 follows from this by R5.

Next, notice that Fig. 3 can be transformed by R6(a) into Fig. 5, which means 'It is possible that not-g is false', that is, 'g is possible'. Hence, from 'g is necessary' (Fig. 2) it follows that 'g is possible' (Fig. 5). Consider the graph of Fig. 6. Since it is the denial of Fig. 5, it must assert that g is impossible.

Fig. 5

Fig. 6

It is now easy to show that 'g is impossible' (Fig. 6) implies 'g is false' (Fig. 8). For from Fig. 6 we obtain Fig. 7 by R6(b), from which Fig. 8 follows by R5.

Fig. 7

Fig. 8

Consider Fig. 2 again. We have seen that it implies Fig. 4.

Fig. 2

g

Fig. 4

Fig. 9

But by R6(a) it also implies the graph of Fig. 9, 'It is possibly false that g is possibly false', or 'It is possible that g is necessary'. Yet, cautions Peirce, Figs. 4 and 9 "can neither of them be inferred from the other" (4.519).

When lines of identity are employed with the broken cut, quantification is determined in the usual way, and for this purpose the broken cut is counted as a standard cut. The following graphs from Ms 468 illustrate this. Note the readings 'may be' and 'can be' for possibility.

It is not required that some a is b, i.e., it may be that no a is b.

It may be that every a is b. (To see that this reading is correct, imagine inserting a standard double cut just inside the broken cut.)

It may be that if a then b *de inesse*, i.e., it may be that a materially implies b.

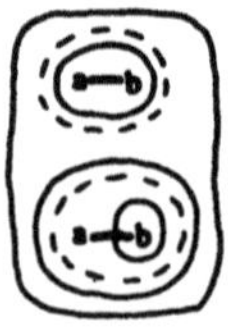

Either no a can be b or any a can be b.

The broken cut may be used with graphs of graphs to state graphical permissions. The spots were introduced earlier in section 5.12.

It is always permitted to scribe a line of identity on SA with its extremities attached to blanks. (The Beta axiom.)

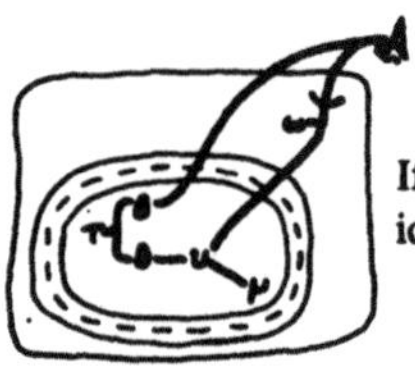

If a dot is on SA it is permissible to attach to it a line of identity with the other extremity at a blank.

The next example is intended to express a principle of excluded middle for graphs.

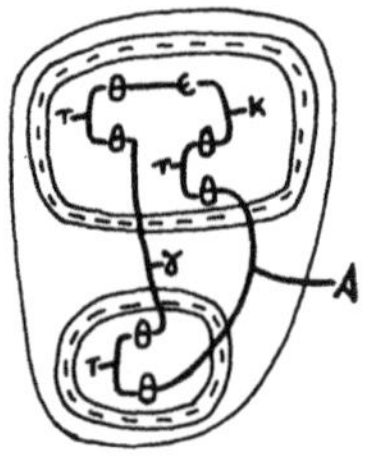

Take any graph, say g; either g may be placed on SA or g may be placed on an enclosure which is on SA.

In C10, stress is placed on the word 'logically'. Peirce emphasizes this by prefixing 'Beta' to modal terms, as in 'Beta-possible' and 'Beta-impossible' (4.516). His meaning is that the broken cut *per se* makes reference to no knowledge other than the knowledge of formal logic embodied in the Alpha and Beta parts of EG (4.515). In particular, no reference is made to the way things are in the sensible world, to the state of the actual universe. This latter is not irrelevant to all uses of modal terms, however, as Peirce proceeds to show.

Suppose that neither the proposition g nor its denial are self-contradictory, and suppose that our state of information is such that g may be true and g may be false. (For example, let g be the proposition 'It will snow

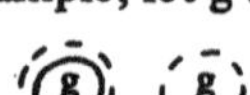

Fig. 11

in Waterloo the day after this book is published'.) We are then entitled to scribe the graph of Fig. 11. Suppose now that we increase our information and learn that g is true. If we add this to Fig. 11, we obtain Fig. 12. But relative to our new state of information, g ceases to be true, and we are thus entitled to scribe Fig. 13. Here we seem to have a derivation from 'g is

Fig. 12

Fig. 13

true' to 'g is necessarily true' (Fig. 13). But we saw earlier that the converse derivation holds, that 'g is necessarily true' (Fig. 2) implies 'g is true' (Fig. 4). If both of these derivations hold, the notion of necessity collapses to the notion of truth, and modal logic would add nothing to the logic of propositions. To prevent this from happening Peirce employs the doctrine that "possibility and necessity are relative to a state of information" (4.517; cf. 2.347), and he introduces a sign to call this to the reader's attention. Namely, he attaches cross marks to the broken cut

> to distinguish the particular state of information to which it refers. And a similar sign has then to be attached to the simple *g*, which refers to the state of information at the time of learning that graph to be true [4.518].

Instead of Fig. 12, then, we have Fig. 14:[14]

Fig. 14

The result is that relative to a state of information there may indeed be a collapse of modal notions into the notions of truth and falsity. But only for omniscience would the collapse be total.

Peirce calls these marks 'selectives', and he points out that they are peculiar in that they refer to states of information as though these were individual objects.[15] They are in fact lines of identity signifying abstractions, like the peculiar lines used with potentials. This becomes clear in the

Fig. 15

14 This corrects Fig. 187 of 4.518. Other graphs in 4.518-522 (Fig. 193 of 4.522 in particular) are not quite as Peirce had them, and they are corrected in the text above.

15 This use of the word 'selective' is not as different from the earlier use introduced in Chapter 4 (end of 4.1) as it might seem. The modal selective, however, is not merely an abbreviation for another symbol.

graph of Fig. 15, which asserts that there is a conceivable state of information in which the knower would know g to be true and yet would not know another graph h to be true. And here, says Peirce, "we have a new kind of ligature, which will follow all the rules of ligatures" (4.521). These selectives have a definite order of succession, and they provide us with an additional rule for EG, the seventh in our list:

R7. The rule of modal selectives. *From* g *we can infer* .

Presumably the number of selectives indicates the order of selection.

R7 is similar to a 'rule of necessitation' [16] which holds for some contemporary modal logics. That rule may be stated as follows: 'If A is a theorem, then necessary-A is a theorem'. Peirce's rule appears to be broader in that the condition is extended to non-logical matters and not restricted to theoremhood in a given system of logic. On the other hand, when the selective is employed in Ms 467 Peirce refrains from reading any graph of the form  by the word 'necessary'.

Presumably, then, the broken cut with its selective expresses in graphical form Peirce's thesis—undergoing serious rethinking in 1903—that "by varying the supposed state of information all the varieties of possibility are obtained" (3.442). This thesis distinguishes two extreme varieties: (1) essential or formal possibility, namely "that which supposes nothing to be known except logical rules"; and (2) substantive possibility, which "supposes a state of omniscience" (3.442; and see 4.67). The broken cut without the selective represents essential possibility; used with the selective, it represents any other variety of possibility, not necessarily excluding substantive possibility, but principally some state of information intermediate between the extremes.

One last note. To express that a state of information B follows after the state of information A, Peirce suggests using a symbol he had employed five years earlier in another connection:

This gives us a symbolic expression of R7:

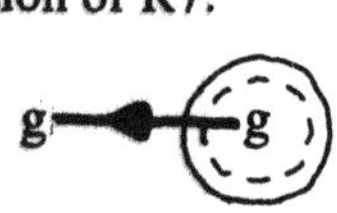

16 So-called in Hughes and Cresswell (1968), for instance. No general comparison of Gamma and the broken cut with contemporary modal logic is intended here. A good beginning, restricted to certain modal propositional calculi, is made in Zeman (1964).

6

TINCTURED EXISTENTIAL GRAPHS

The 1903 account of Gamma contained some fabulous ideas, but Peirce was not satisfied with it. In the first place, he was sure that there were rules of inference for Gamma yet to be discovered, although he thought he had given "perfect rules" for the broken cut. In the second place, he had been unable to develop the purely syntactical exposition of Gamma that he had aimed for; the use of his new symbols depended too heavily on their significations. He said this much in the 1903 *Syllabus* (Ms478, p. 157).

These problems were never solved by Peirce, but he avoided them by working up a brand new treatment of EG in which no special Gamma symbols would be required.[1] In fact, in its final form, the new exposition would not even distinguish the three parts Alpha, Beta, and Gamma, although all the elements of Alpha and Beta would be retained.

It is important to emphasize at the very beginning of this chapter that Peirce made absolutely no changes in the graphical rules of inference while he was producing this exposition.[2] The rules he employed throughout are essentially the same as those given in Chapter 4 above. For this reason the

1 Peirce continued to use certain of the potentials and graphs of graphs after 1903. Potentials crop up in notebooks, loose sheets, and even letters to William James. By 1908 he was using Roman numerals for potentials, as in 4.620ff. In a March, 1906 entry of the *Logic Notebook*, p. 273r, he employs some old and some new graphs of graphs.

2 Perhaps 'absolutely' is just the slightest bit too strong. He did add the new phrase "in different Provinces" to one old rule (Fourth Permission, 4.569) which had appeared as a derived rule in many expositions of EG since 1898; it can indeed be derived from our Chapter 4 rules, and Peirce remembered that it was not independent (4.567). The fourth permission is a grahical version of the principles of the distribution of the universal quantifier over conjunction and of the existential quantifier over alternation. Peirce's scepticism regarding its validity (Ms L 477, quoted in CP 8, p. 298; cf. a letter to William James dated February 26, 1909, p. 23, in Ms L 224) is not justified. He also made a slip in stating the second permission which he detected and pointed out to Lady Welby in a letter written in February, 1909 (Ms L 463, verso of p. 4). The last sentence of the rule in question, "This involves the Permission to distort a line of Identity, at will", seemed to permit the joining of two evenly enclosed lines.

rules are not discussed until the last section of the chapter, when a few suggestions for changes and additions will be made. Peirce himself suspected that modifications might be necessary (Ms 490 at the 4.581 ellipsis), but he apparently did not pursue his new discoveries long enough to find them.

6.1 THE TINCTURES

The germ of the new exposition was present in 1903, when it had occurred to Peirce to replace the single sheet of assertion with the book of separate sheets in order to deal with logical possibilities. As pointed out in the last chapter (section 2), the idea did not seem convenient at the time, and he worked instead with the broken cut. Further elaboration of the idea was not reported until the spring of 1906, when Peirce announced his "very recent discovery" (4.576) that the area within the cut, to be viewed as the verso of SA, represents a kind of possibility.

> The cut may be imagined to extend down to one or another depth into the paper, so that the overturning of the piece cut out may expose one stratum or another, these being distinguished by their tints; the different tints representing different kinds of possibility [4.578].

The cut would retain its function of negation, but the effect of scribing a graph on the verso would be to *exclude a possibility* (not simply an actuality) from the universe.

Peirce arrived at this analysis while considering the "anomaly" by which in EG the relation 'other than' is expressed differently from any other relation (*LN* 265r). It is the only relation requiring that a graph be partly in one area and partly in another; it requires that a line of identity cross a cut. Thus, the graph of Fig. 1 asserts that some woman is other than any angel. Hitherto,

Fig. 1

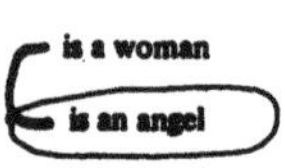

Peirce would have said that Fig. 1 contains two lines, one on each area, and joined at the point on the cut. But now he observes that the graph does not assert that there is any such *thing* as an angel, but asserts that there is an *idea* of an angel. It means, says Peirce, that "she lacks some essential character of an angel" (*LN* 266r). This form of proposition, then, does not relate one existent individual to another, but relates an existent to a character, that is, to a possibility. Such a relation, between subjects of different categories of being, Peirce called a 'reference' (3.572). And it gives him an interpretation for a graph that crosses a cut. The investigation had begun on December 14,

1905 (*LN* 264r), and on March 9, 1906 (*LN* 274r) Peirce concluded that all discourse admits of but two kinds of objects: actualities (represented on the recto of SA) and possibilities (represented on the verso).

When Peirce announced his discovery at the April meeting of the National Academy of Science, he was hopeful that it would singlehandedly enable him to complete Gamma. It did not. The new interpretation did enable him to symbolize references in EG; and it did enable him to express a conditional proposition which would not be true simply because its antecedent failed. Thus, on this new interpretation, Fig. 2 asserts 'It is not possible that

Fig. 2

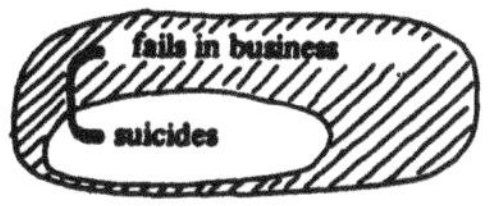

a man fails in business without suiciding'. The shading represents Peirce's blue tint (done with a blue pencil; see Ms 490, the passage omitted at the end of 4.575). This is more than a conditional *de inesse* (the material conditional); it is in fact a strict implication (more is said about this issue in section 2). But the cost of these two benefits was too high. By making the cut do double duty, there was no way to separate the expression of a possibility from the expression of a denial. The inability simply to affirm a possibility had to be remedied, not only for purely logical reasons, but because Peirce was becoming more and more insistent on recognizing in a formal way his belief that there is objective possibility, a possibility which is not based on ignorance. The inability to express simple denial had to be remedied because without it, the conditional *de inesse* could not be diagramed. Although Peirce was currently modifying his former staunch Philonian viewpoint, he continued to admit "with Scotus" (Ms 292, p. 51) that there is a conditional *de inesse.*

The shortcomings were not noticed immediately. They gradually came to light in April and May, as Peirce tried to work his discovery into a full blown exposition of the graphs, to be included in his third *Monist* article on pragmaticism. His earliest drafts of that article, in Ms 292, still recognize only the two universes of actuality and possibility, but they contain first mention of the 'tinctures', the patterns suggested by heraldry which Peirce would soon use to distinguish surfaces from each other. The heraldic tinctures were originally invented to solve the problem of how to represent colors using only black and white. Peirce simply adapted them to suit his own purpose.

The first improvement on this schema, one which brings it into line with Peirce's doctrine of the categories,[3] is the addition of a third universe: the

3 It is something of an impertinence to treat the categories as briefly and as indirectly as I do, but the full story of the relationship between the categories and EG is beyond the scope of the present book. I must assume that the following description

universe of Destined Results, or whatever "cannot be considered as affirming or as denying either an actuality or a possibility" (Ms 295, p. 43). The recto will be symbolized by the heraldic tinctures of metal, the verso by the tinctures of color. The third universe will be represented by patches of fur, to be "sewn" on the recto or verso as required. The names of the universes vary somewhat from draft to draft: the recto denotes the Actual and the Existent, or the Actual and True, or Actual Fact; the verso denotes the Questionable and the Possible, or things Possible in themselves; and the aggregate of patches of fur denotes the Commanded and the Compelled, or what is sure to be (Ms 295).

A second improvement on the early scheme is a partitioning of the three universes into ten sub-universes. These latter universes are designated by specific tinctures, that is, by patterns like those of Fig. 1 in section 2 below (the partitioning in 6.2 differs from that to be given here, and the difference will give you an idea of the course of Peirce's thought; cf. 4.553 n.1). In one early list of Ms 295 there are two metals: (1) *argent* refers to the actual or true in a general or ordinary sense; (2) *or* refers to the actual in some special sense. There are five colors: (1) *azure* refers to logical possibility; (2) *sable* refers to subjective possibility (that which is not known to be false); (3) *gules* refers to a more objective mode of possibility; (4) *purpure* refers to ability; (5) *vert* refers to what is in an interrogative mode. There are three furs: (1) *vair* refers to what is commanded; (2) *potent-counter-potent* refers to the compelled; and (3) *ermine* refers to the rationally (or metaphysically or secondarily) necessitated.

A heavy dot or line on SA has, till now, denoted an existing individual. With the increase in universes and sub-universes, however, its interpretation will depend upon the tincture of the area on which it is scribed. If scribed on a metal, it will still refer to an existing object of the universe of actuality; if scribed on color or on fur, it will refer to an abstraction of whatever sort the specific tincture represents. A special difficulty arises in the case of a line of identity which extends from one tinctured surface to another on the same area, and we must decide what sort of being that line will denote. A method for doing this will be presented later. Here I simply point out that such a line will enable Peirce to relate actualities to possibilities without requiring the line to cross a cut, that is, without requiring a graph to be partly in one area and partly in another. See the discussion of Fig. 1 above.

A third improvement remedies the second mentioned shortcoming of the

---

of 1903 will remind the reader of the doctrine: The categories are the three modes of being which can be directly observed in whatever is before the mind in any way at any time. They are Firstness, the being of positive qualitative possibility; Secondness, the being of actual fact; and Thirdness, the being of law that will govern facts in the future (1.23).

April lecture, the inability to express simple denial. Thus far denial had involved scribing a graph on the verso, making a cut around the graph, and overturning the excised piece to expose it. The result would be the exclusion of whatever is scribed on the overturned area from whatever kind of possibility the area represents. (If the graph to be negated contains a cut, the twice negated graph within that cut must be scribed on the recto, and so forth.) This version of denial will be retained, but in addition the Graphist is now permitted to make the cut and leave the excised piece where it is, without overturning it. This results in a denial having the mode of truth belonging to the place of the cut. Thus, in Fig. 3, the cut is made on the

Fig. 3

Fig. 4

recto of SA and the piece is overturned bringing the verso into view; the graph means 'It cannot rain'. In Fig. 4, the graph is scribed on the recto and the incision is made, but the piece is not disturbed; the meaning is 'It does not rain'. The cut in Fig. 3 has a verso area, that in Fig. 4 a recto area. To mark this distinction Peirce introduced a finely dotted line to represent a cut having a recto area, while the standard oval line represents a cut having a verso area.[4] Use of the dotted cut for the graph of Fig. 4 produces Fig. 5.

Fig. 5

A fourth, and relatively late, improvement in Ms 295 remedies the first mentioned shortcoming of the April lecture, making the system capable at last of affirming (not just denying, or excluding) possibilities. The single sheet whose recto represents actuality and whose verso represents possibility is replaced by a 'Phemic sheet' whose signification depends upon the tincture of its outermost border or rim. In Peirce's terminology, a pheme is a grammatical sentence, whether interrogative, imperative, or declarative. Peirce means to allow for the expression of all such varieties of sentences in his new version of EG, and the term 'Phemic sheet' calls attention to this.

When the outermost border of the Phemic sheet is metal, usually argent (symbolized by the blank, white sheet), the sheet will be called the *Sheet of Assertion* and will be devoted to the expression of propositions. When the border is in color, usually azure, it will be called the *Sheet of Interrogation* and will be devoted to the expression of questions. When the border is in fur, usually sable (which has moved over from color), it will be called the

4 Ms 295, pp. 44, 47, 81. It is this dotted cut that occurs in the figures of 4.564-571. Do not confuse it with the dotted rim or the broken cut of Chapter 5.

Sheet of Destination and will be devoted to the expression of resolutions. It is stated that every part of the sheet will have some tincture or other, and the graphist may tincture the various parts according to his needs. If for example he wishes to express on the sheet of assertion (metal) the proposition 'It is possible that it will rain', he must tincture some part of the sheet with a tincture of color, say azure, and scribe 'It will rain' on that part. The graph is given in Fig. 6, where the rectangle is used simply to mark the edge

Fig. 6

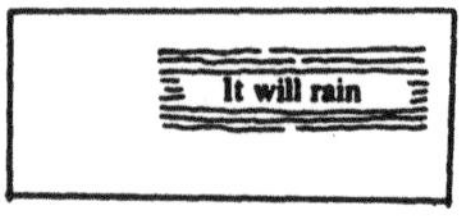

of the Phemic sheet so that the border in argent–represented by the white of the paper–can be detected. You may assume that all graphs scribed in the rest of this book are scribed on metal unless otherwise noted; and I will assume that no further rectangles are necessary.

In these later drafts Peirce emphasized a notion he had spelled out for EG as early as 1898 (Ms 484) and again in 1903 (4.431). He urged the reader to imagine that the graphical analysis is a collaboration between two parties, whether they are two groups of people, two people only, or two mental attitudes or states of one person. That the two parties are present in one person follows from the view that "Reasoning is nothing but the discourse of the mind to its future self" (Ms 450, p. 3). And the purpose of existentional graphs is "to aid one in talking to himself" (Ms 650).

One of the parties, called the Graphist, is responsible for scribing the original graphs at the beginning of the investigation or discussion; the other, called the Interpreter, draws inferences from these graphs by changing them in accordance with the permissions of the system. The Phemic sheet, before anything is scribed on it, represents whatever is taken for granted at the outset by the Graphist and Interpreter.

6.2 PROLEGOMENA

The final result of the development I have sketched, Peirce's last full scale revision of EG, was published in an article for the October 1906 *Monist*: "Prolegomena to an Apology for Pragmaticism" (4.530-572). All the basic elements have been described already, and there is only one major refinement to be reported. It is this: the Phemic sheet is no longer the only sheet we are working with. Rather, we are to imagine having an abundant supply of different sheets, each with its own tincture and each with its own recto and verso. Although the verso is supposed to be a rougher surface than the recto, the two sides of a sheet will have the same tincture; thus the three-step process of negation can be reinstated without qualification, since there is no danger of undoing the effect of the third improvement. At the start of any investigation, the Graphist and the Interpreter designate one of these sheets as the Phemic sheet. The other sheets are to be inserted into or placed upon the Phemic sheet as needed. The whole image is that of patchwork: the sheets are cut up and pieced together on the Phemic sheet in the very scribing of graphs.

There are two minor changes regarding the tinctures. First, the third mode of being is now called 'intention'; but the representation of actuality by metal, possibility by color, and the third universe by fur is unchanged. Second, each mode of tincture is a class or collection of *four* tinctures. Why four?

Different states of things may all be Actual and yet not Actual *together*; and the same is true of the Possible and the Destined. Two graphs in the same *Province*, i.e. on the same continuously tinctured surface will be asserted, not merely as True, but as True *together*. Hence, since four tinctures are necessary to break the continuity between any two parts of any ordinary surface, four metals, four colors, and four furs will be required [Ms 295, p. 44].

Note the definition of 'province' as 'a continuously tinctured surface'.

Fig. 1 below contains the twelve tinctures as Peirce gave them in Prolegomena. Do not allow yourself to be put off by the difficulties you would face in trying to use the printed patterns; they were meant for publication only. In his own practice Peirce used colored ink and colored pencils; pencils today are easy enough to obtain in at least sixty different hues. It is doubtful that as many as sixty would be necessary for any mortal philosopher. In any case, the visible spectrum, with variations in brilliance and saturation, affords all the differences that could ever be required. We might then simply "fancy or pretend" (Ms 300, p. 39) that there is some similarity between the three fundamental hues and the three Universes. However, in 1906 reproduction of colors was "in print impracticable", so Peirce resorted to the heraldic tinctures for Prolegomena (Ms 300, p. 40).

In order to make the illustrations which follow as clear as possible, and to make it easier for the reader to begin experimentation of his own, I include in Fig. 1 an indication of the universes that might be represented by the tinctures and the pencil colors which have been convenient in practice. I have not yet found a use for fer or plomb.

We begin our illustrations by considering Fig. 6 at the end of the preceding section. In the earlier system of graphs there described, the azure tincture was supposedly applied directly to the Phemic sheet, thus providing for the expression of a possibility. If we imagine instead that an azure tinctured sheet has been placed on the Phemic sheet to produce that same figure,

Fig. 1

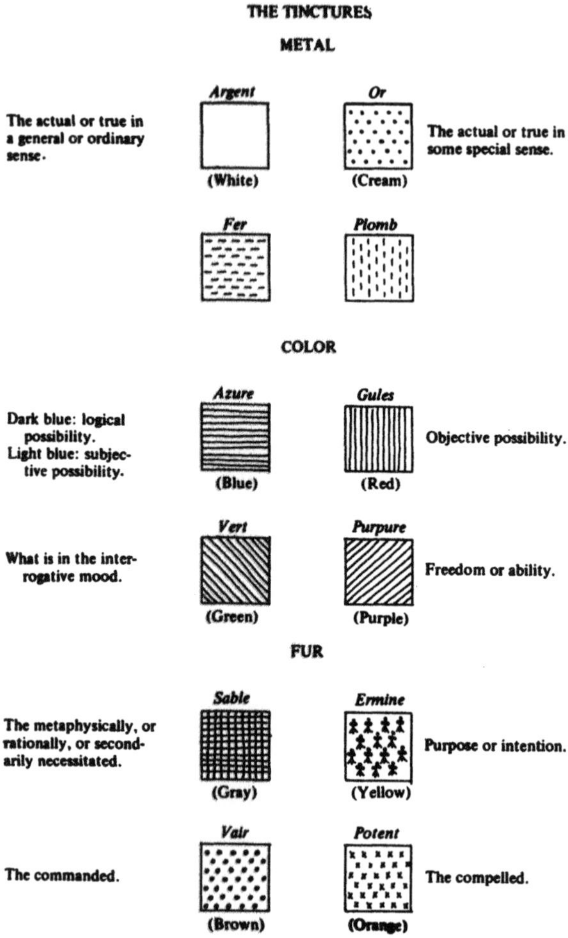

then in the system of Prolegomena it will express the same proposition, 'It is possible that it will rain'.

As a second example we present the graph of Fig. 2 which is taken from

Fig. 2

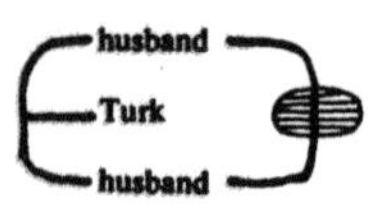

Prolegomena (4.569). This graph asserts that there is a Turk who is the husband of two different persons. The area of the enclosure is on color (azure), and it thus denies the possibility that the individuals denoted by the lines above and below the cut are identical. But would it make a difference if the area of the cut had been on metal, say argent? No, says Peirce of a similar graph in Ms 295 (p. 53). The point is that azure represents subjective possibility, so that Fig. 2 merely says that it is contrary to what is known by the Graphist that the two individuals should be the same. If the area of the cut were on argent, the meaning would be unchanged; for the Graphist still "would give his word" that as far as he knows, the individuals are not the same.

This is not a particularly helpful example of the use of the tinctures, and a reader of Prolegomena will be astonished that it is the only example. There is in the article one line of reasoning in particular which seems to require a tinctured graph and seems unfinished without it. I refer to an argument Peirce developed to support his two related claims, that "there are real possibilities" (4.547) and that the tinctures (or something like them) are necessary.[5]

6.2 1 *On Behalf of the Tinctures*

The argument consists in a logical analysis of the following proposition, which we shall refer to as *Ppn P* (for 'proposition P'): 'There is some married woman who will commit suicide in case her husband fails in business'. The issue revolves around what this proposition means and how it should be represented in EG.

6.2 11 Peirce distinguished two interpretations of *Ppn P*. The first, which we shall call *P-de inesse*, takes *Ppn P* to mean only that there is a married woman, and either her husband does not fail or she commits suicide (4.569). The word 'only' emphasizes that *P-de inesse* does not assert that the failure has any connection with the suicide (see *LN* p. 320r).

This is the interpretation of logicians for whom every conditional is a Philonian (material) conditional.[6] Peirce, at one time "decidedly" of this opinion (Ms 292, p. 51), reversed himself when it became clear to him that the Philonian logician could "take no consistent position other than that unactualized possibility is unreal" (*Ibid.*). His realism and his pragmatism had convinced him of the reality of possibility. Common sense too: "I find after all that I have no doubt that I really *can* raise my arm whether I actually do so or not" (Ms 292, p. 52; cf. 4.579). He therefore could no longer accept the Philonian version as the interpretation of every conditional, although, as was made clear in 6.1, he continued to acknowledge that there are such propositions. Perhaps *Ppn P* is one of them?

To find out, we consider the graph of *P-de inesse*, given in Fig. 3. Now the trouble with this proposition is that its truth is too easily guaranteed.

5 The argument is spread out, but 4.546 and 549 contain what is essential for our purpose. A parallel argument appears in 4.580 (Ms 490).

6 See for example 3.442-443, published in 1896. Also Ms 292, p. 51.

For it will be true if either some married man does not fail in business (whether his wife commits suicide or not) or some married woman commits suicide (whether her husband fails in business or not). To show this, let us

Fig. 3

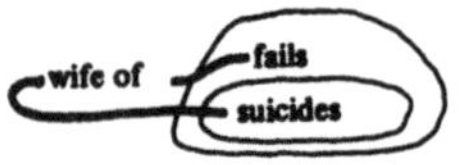

suppose the first alternative is true.[7] Then we can scribe 'Some married man does not fail in business' on SA (an abbreviation I will continue to use for the metal Phemic sheet), as in Fig. 4. From this graph, by R2 (insertion onto

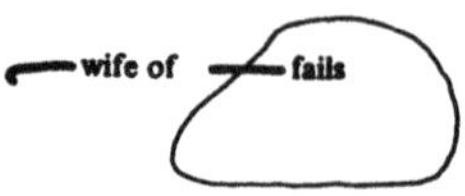

Fig. 4

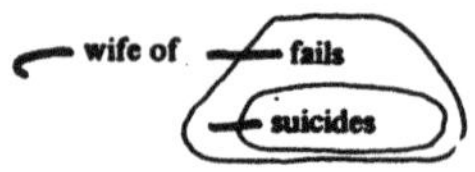

Fig. 5

odd), we obtain Fig. 5. By R3(b) the unenclosed line can be extended into the first area of the scroll of Fig. 5, after which joining the loose ends by R2 yields Fig. 3. This is satisfactory as far as *P-de inesse* is concerned, but to say that *Ppn P* is true just because there exists a married man who does not fail, is absurd. Hence *P-de inesse* is an inadequate interpretation of *Ppn P*.[8]

6.2 12 Instead, says Peirce, "what is really meant" by *Ppn P* is that there is some married woman who *under all possible conditions* would commit suicide if her husband fails in business (4.546). We shall call this proposition *P-modal*. It would now be useful to examine a graph of *P-modal*, especially since the point of adding the tinctures to EG was to render such propositions expressible. No such graph appears in Prolegomena, which is rather curious since in the April lecture (see 4.580) Peirce had already scribed graphs of the required form. But that lecture represents the earliest stage in the development of the tinctures, the stage in which possibility and negation could not be separately expressed. The manuscripts written at that time contain many graphs illustrating this early stage of development, but almost none (perhaps only two) of the final stage which is reported in Prolegomena. It is quite likely that the final improvements came in such a rush that Peirce had insufficient time to 'get over' the early system and 'get used to' the new one.

7 Let the reader construct a similar proof beginning with the second alternative 'Some married woman commits suicide'.

8 Peirce formerly held that the inconveniences of the Philonian conditional, nowadays known as the paradoxes of material implication, could always be overcome by combinations of Philonian conditionals and denials of conditionals (3.443).

Nevertheless, we can supply what is missing and complete the analysis ourselves. Figs. 6 and 7 both express *P-modal*. The shaded areas are on

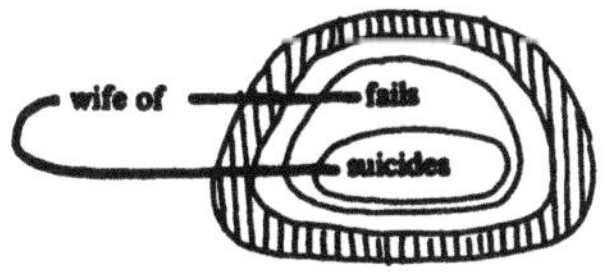

Fig. 6

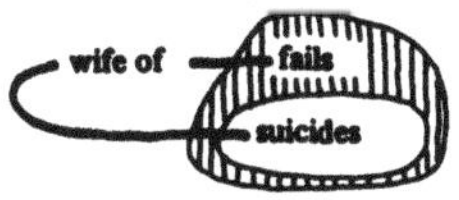

Fig. 7

gules to represent objective possibility. Fig. 6 means 'There is a married couple, and it is necessarily the case that if the husband fails then the wife suicides'. The shaded ring may also be read 'not possibly not'. Fig. 7 means 'There is a married couple, and it is not possibly the case that the husband fails while the wife does not commit suicide'. That these two readings are equivalent is essentially a matter of definition, although it may not be immediately apparent from the graphs. (The graphical equivalence will be demonstrated in the next section of this chapter.) In *Principia* notation, using the square and the diamond for the necessity and possibility operators respectively, the propositions of Figs. 6 and 7 look like this:

Fig. 6 $(\exists x)(\exists y)[Wxy \,\&\, \Box[Fy \supset Sx]]$

Fig. 7 $(\exists x)(\exists y)[Wxy \,\&\, {\sim}\Diamond[Fy \,\&\, {\sim}Sx]]$

The equivalence will now be obvious if it is remembered that, for any propositions A and B, ${\sim}\Diamond A \equiv \Box{\sim}A$, and ${\sim}[A \,\&\, {\sim}B] \equiv [A \supset B]$.

P-modal does not have the disadvantage of *P-de inesse* that was pointed out above. The fact that some married man does not fail in business (Fig. 4) is not sufficient to make *P-modal* true (Figs. 6 and 7). One other difference is worth noting. *P-de inesse* is false of a given married couple if and only if the husband does in fact fail and the wife does not then commit suicide. In order to make *P-modal* false of a given married couple, however, it is not necessary for the husband actually to fail; "it will suffice that there are *possible* circumstances under which he would fail, while yet his wife would not commit suicide" (4.546). This difference is graphically expressed in Figs. 8 and 9:

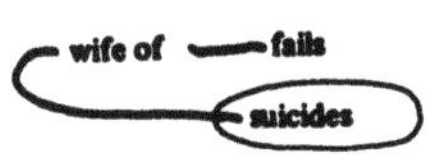

Fig. 8

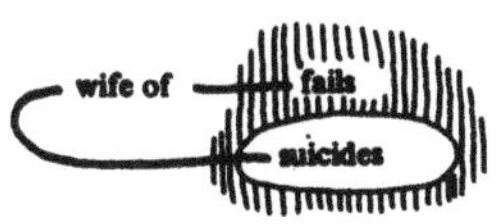

Fig. 9

inspection should convince you that if Fig. 8 is true, Fig. 3 must be false; and if Fig. 9 is true, Fig. 7 must be false.

It is useful to have a means of expressing the difference between *P-de inesse* and *P-modal*, but if EG is really to be adequate for modal logic it must contain rules of inference which would permit us to express such relationships as the equivalence between Figs. 6 and 7, or the inference from Fig. 6 to Fig. 3 (since what is true of all possible worlds must be true of the actual world), or the inference from Fig. 8 to Fig. 9 (since what is actual is possible). No such rules appear in Prolegomena, in its early drafts, or, to the best of my knowledge, in the later papers of Peirce. A first attempt to provide such rules will be made in the next section of this chapter. In the remaining pages of this section we will examine the fate of the tinctures after Prolegomena.

6.22 *1906 and After*

In the later drafts of Prolegomena Peirce admitted that he had tried only four tinctures: argent, or, azure, and gules. His study of even these was "by no means" thorough, and he was far from certain that they would "work smoothly" (Ms 295, p. 53). He confessed this, he said, in order "to avoid leading other students astray", that is, in order to make it clear that these parts of the system were in the experimental stage. That the tinctures were still in the experimental stage in the published article is partly explained by the fact that Peirce allowed himself no more than four months to write it. As late as January 8, 1906, he did not intend to include an exposition of EG in his third *Monist* article on pragmaticism, and by May 29 the editor of *The Monist* had Prolegomena in his hands.[9] Continued experiment with the tinctures was apparently limited by the urgency of other work (see 4.653 for a typical comment), some of it directly related to, and even precipitated by, his new version of EG. In the first place, there were two more articles in the series and their promised proof of pragmaticism to produce. In the second place, the tinctures seem to mark Peirce's attempt to begin a systematic investigation of the sort he had called for in his 1903 *Syllabus* of the Lowell Lectures, when he despaired of completing Gamma:

> The Gamma part of the system of Graphs can never be perfected until we have precisely analyzed all the conceptions of logic in terms of the three Categories, together with such other exact conceptions as it may be found necessary to add to those of the Categories. But this is a labor for generations of analysts, not for one [Ms 478, pp. 164-165].

Peirce kept looking for improvements in EG along the lines of Prolegomena, for several years keeping the scope of his graphs wide enough to include

9 The January date occurs in Ms 283, p. 56 (5.554), a draft of the third article in which the graphs are mentioned only in passing. The May date is from a letter of Paul Carus to Peirce, Ms L 77. I am indebted to Max Fisch for calling the first passage to my attention and for supplying the May date, which was particularly important in establishing the order of Peirce's discoveries.

"creations of explanatory conjectures, as well as the whole process of induction".[10] But the tinctures remained somewhat "perplexing" to use,[11] and after the spring of 1908 they were not often mentioned. Brief remarks about EG in the April *Monist* of 1908 include mention of Prolegomena but not of the tinctures (4.617); some accounts in letters and unpublished papers also ignore them,[12] a few are critical of them.[13]

There are still noteworthy passages, however, which reveal something about the course of Peirce's later thought. We will consider five such passages.

1. In an undated manuscript (for which early in 1908 might be a good guess) Peirce claims to have made a "slight" improvement on Prolegomena: he suggests drawing (preferably with a red pencil) a border around SA, and "in the margin outside the red line, whatever is scribed is merely asserted to be possible" (Ms 514, pp. 18-19). This "improvement", reminiscent of the dotted oval of 1903,[14] does not become a permanent feature of EG.

2. In September of 1908 Peirce reconsiders key propositions from Prolegomena, and scribes the graphs of Figs. 6 and 7 above in his *Logic Notebook* (p. 320r). He misinterprets Fig. 7, reading it as though the shading applied only to the spot 'fails'. Other similar propositions are also translated into his algebraic notation. To account for necessity in this notation he introduces an index (the letter omega) to denote "a state of things", and he applies the universal quantifier to it. The resulting form, Π_ω, is read "under all circumstances", and it is immediately employed to express the proposition *P-modal* (*LN* p. 319r):

$$\Sigma_i \Sigma_j \Pi_\omega \; w_{\omega ji} \cdot (\bar{f}_{\omega j} \Psi s_{\omega i}).$$

Four months later this way of expressing necessity is adapted to the graphs. The instrument is a line of identity which looks and functions suspiciously like the modal selective of 1903 (5.2. above). This line is attached to the tops of graphs to which it applies, and it obeys the usual graphical conventions for quantification. Among the examples given (*LN* p. 340r) are these three:

10 This is from Ms 296, p. 7, written about March of 1908.

11 Ms 300, p. 40, written about March of 1908. He had recognized from the first that there was difficulty involved in the use of the tinctures. (Ms 292, two pages numbered 45.)

12 As in Ms 514 of 1909, a letter to Lady Welby of January and February 1909 (Ms L 463), a letter to William James of February 26, 1909 (or begun then), and Ms 650 of July and August 1910.

13 As in a letter to Allan Douglas Risteen of December 1911 (Ms L 376), and a letter to Frederick Adams Woods begun in October 1913 (Ms L 477).

14 See the beginning of 5.13 above, where Convention No. 14 from *Logical Tracts. No. 2* is discussed.

p is true under some circumstances.

p is true under all circumstances.

Under all circumstances somebody wins & somebody loses.

Using this notation it is easy to scribe *P-modal* in EG without using the tinctures (the graph of Fig. 10 does not appear in *LN* or anywhere else to the best of my knowledge, but it is quite possible that Peirce had it in mind).

Fig. 10

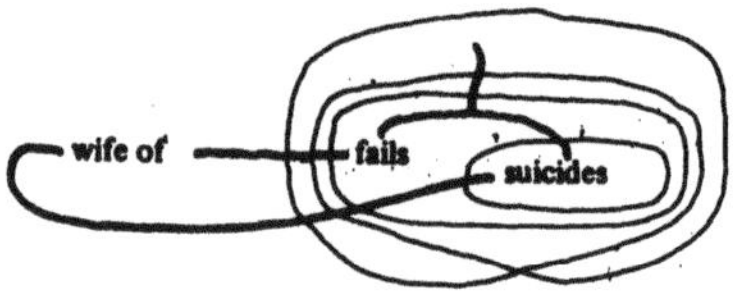

It would have been difficult for Peirce to be satisfied with this. For one thing, the tinctures were designed with more than formal logic in mind; they were meant to provide a structure in terms of which Peirce could apply his categories to propositions and inferences, to hypotheses, questions, and commands, to "all that ever could be present to the mind in any way or any sense" (Ms 499(s)) – and to do this with the same system that handled the formal logic as well. Indeed, once the tinctures were set up and in running order, a steady use of the graphs could be expected to turn up discoveries that would enrich and modify the theory of the categories, while continuing study of the categories would improve the classification of the tinctures. The 'modal selective' just illustrated would lend itself to none of this.

Secondly, this modal selective is essentially another 'peculiar' line of identity, of the same order as those introduced along with the 1903 potentials and graphs of graphs. But since 1903 Peirce had become convinced that special symbols of this sort were inappropriate for the task he had in mind. He made this clear in a manuscript written early in 1908, when he reviewed the history of his tinctured graphs. His studies had convinced him of the need to further differentiate the categories, and so he distinguished the sub-universes which the tinctures represented. The tinctures seemed necessary because the differences in universes and sub-universes

> are not differences of the *predicates*, or *significations*, of the graphs, but of the predetermined objects to which the graphs are intended to refer. Consequently, the Iconic idea of the System, requires that they should be represented, not by differentiations of the Graphs themselves but by appropriate visible characters of the surfaces upon which the Graphs are marked [Ms 300, pp. 38-39].

This, I think, explains why the broken cut was abandoned after 1903.

Thirdly, the development of the tinctures was concurrent with the development of Peirce's notions of possibility and necessity. Summarizing these changes in the same manuscript just quoted, Peirce said that he had come to view necessity as "that which tends to govern both Thought and real Fact, even should it never become absolute in either sway" (Ms 300, pp. 39-40). Now it is likely that the new and broader view does not require a total rejection of the earlier view of necessity as "that which is true under all possible circumstances",[15] but it is the earlier view that the graphs of January 1909 express, and that is all they express.

Sure enough, on February 26 of 1909, the reality of possibility asserted itself again and Peirce penned the following "Note on the Tinctures" in his *Logic Notebook* (p. 345r):

To illustrate the need of the Tinctures, take the fact that no matter what Particles there may be on a Line, there will be a point place on that Line (if nothing but Particles are on it) where there is no Particle but where a particle can be placed.

3. A third noteworthy passage from the late papers, this one written on September 13, 1910, occurs in what Peirce called '*The Prescott Book*'. "It is my duty to investigate Modality more closely", he wrote, and he scribed a graph expressing that a person may dream that Theodore Roosevelt attacks him (Ms 277, p. 171):

Fig. 11

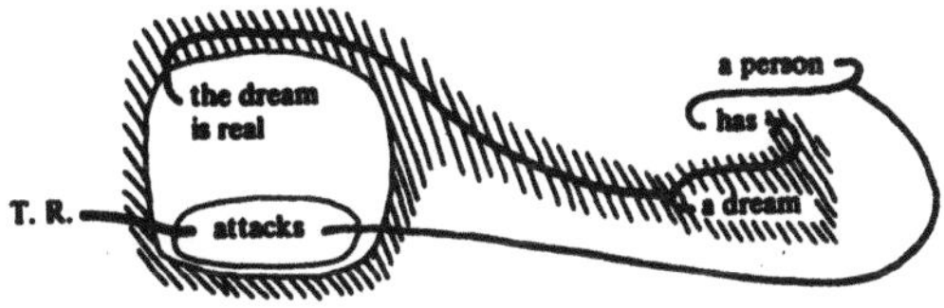

The shaded area represents the "field of may-be".

4. The tinctures are included in an exposition of EG written in June of 1911.[16] There the graph of Fig. 12 is scribed to express the proposition

Fig. 12

15 Ms 300, p. 38. See also 6.590ff written in 1891.

16 The manuscript is Ms 670, "Assurance through Reasoning". It is a revised version of Ms 669, which was begun in May of 1911. Peirce hoped to contribute this paper to a book of essays edited by Lady Welby and others. See Peirce (1953), 42-46.

'Either there is a mind or there is an absurdity'. But this does not quite assert that the existence of a mind is an absolute necessity, says Peirce, since it does not say that there *would* be a mind under all possible circumstances. To make this explicit, it is necessary to indicate on the border of the Phemic sheet that the logical universe is that of metaphysical possibility.

This would seem to be an extension of the significance of the border tincture. Previously, in Ms 295 and in Prolegomena itself (see the very end of section 1 above, and 4.554), there were three choices only for the Phemic sheet: if its border was tinctured in any metal, it was to be taken in the indicative mood, as expressing propositions; if its border was in any color, it was to be taken in the interrogative mood, as expressing questions; and if its border was in any fur, in the imperative mood, as expressing resolutions. With the present improvement, the sheet will presumably represent whatever sub-universe its border tincture represents. To indicate that whatever is scribed expresses a question, not just any color will do; it will be necessary to use some particular color – on our scheme, vert. The universe of metaphysical possibility might be denoted by gules. In this way, there is no limit to the universes that the Phemic sheet can represent.

After the above comment regarding Fig. 12, Peirce continues as follows:

The nature of the universe or universes of discourse (for several may be referred to in a single assertion) in the rather unusual cases in which such precision is required, is denoted either by using modifications of the heraldic tinctures, marked in something like the usual manner in pale ink upon the surface, or by scribing the graphs in colored inks [Ms 670, pp. 18-19].

When tinctures are used, then, as in Prolegomena, the metals are used to mark different kinds of existence or actuality, the colors mark different kinds of possibility (here listed as "possibility consisting of ignorance, of variety, of power, of futurity"), and the furs mark different kinds of intention. When what is scribed is not intended for publication, however, "nothing else is so simple as the use of colored inks". (Except perhaps the use of colored pencils.) To express the metaphysical necessity of mind, Peirce instructs us to scribe the graph of Fig. 13, which reads, literally, "It is metaphysically impossible that there should not actually exist a mind" (Ms 670, p. 20).

Fig. 13

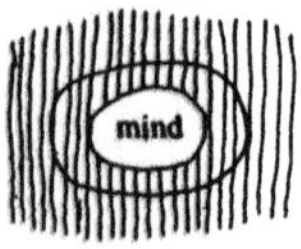

In this manuscript (and also in Ms 674) there is interesting evidence that Peirce had been following out the 'program' suggested by the tinctures, namely, the analysis of the conceptions of logic in terms of the categories. For he makes use of the notion of command–'the commanded' is a universe represented by one of the furs (vair)–to analyze the notion of assertion.

5. The last passage we examine occurs in a letter Peirce started to write to Frederick Adams Woods in October of 1913. I quote it with one substitution, two additions, and three subtractions: (1) a reference to pages in the October 1906 *Monist* has been replaced by the corresponding reference to *CP* (taken from Burks's bibliography in *CP* 8, p. 297); (2) a figure number has been added to each graph to facilitate our discussion; (3) instead of shading the three oddly enclosed areas of the graphs (one in the first, two in the second) as Peirce did, I use metal throughout, because the shading in the letter had nothing to do with the tinctures (as the letter makes clear) and was employed solely to render the graphs more perspicuous (see 4.617). The interruptions were recorded by Peirce.

[In Prolegomena] I made a blunder quite like that of some over-fatigued computor who makes 9 + 7 = 2 and when he reviews his work repeats the same idiocy. My blunder is contained in [4.569, from "For the sake of illustrating this . . ." up to the statement of the fourth permission]. Curiously enough it was not until two years after the publication that I happened to notice my fallacy.

(Interruption)

Instead of scribing

[Fig. 3]

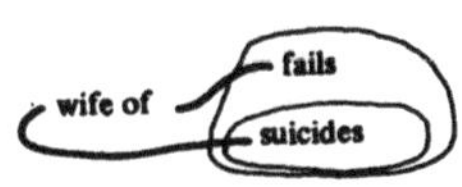

as I did, I should have scribed

[Fig. 14]

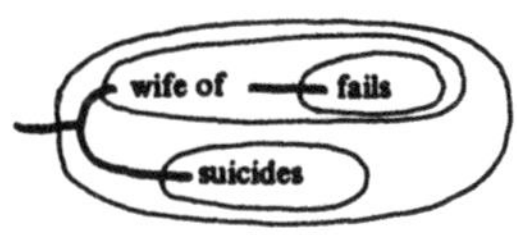

It would pay any very accurate thinker, such as you and Royce are, to examine the fallacy.

(Interruption)

It cost me the trouble of my nonsensical "tinctures" and heraldry [Ms L 477].

Three questions arise: (1) What is the fallacy of 4.569? (2) How would the substitution of Fig. 14 for Fig. 3 repair the fallacy? (3) How did the fallacy lead Peirce to his 'nonsensical' tinctures?

(1) I can find no fallacy in the disputed passage. The passage contains eight sentences, and only one of them seems to me to contain an error. That

one is the fifth sentence, which claims that Fig. 3, *because* it is scribed all in one province, thereby asserts that the wife's suicide is "connected with" the husband's failure. It is not clear what 'connected' means, and the fifth convention (4.562) which deals with the connections of graph-instances does not make it clear. But if the claim is that this undefined connection makes Fig. 3 express more than *P-de inesse*, so that the tinctures are not necessary after all, then the claim must be rejected. For the inference from Fig. 4 to Fig. 3 (above) is valid whatever the fifth convention asserts, and it is the inference which shows both that Fig. 3 precisely expresses *P-de inesse*, and that *P-de inesse* is not an adequate interpretation of *Ppn P*.

This 'error', however, can hardly be the fallacy that Peirce was referring to, since the correction serves to establish, rather than to undermine, the need for the tinctures.

(2) The substitution of Fig. 14 for Fig. 3 in the disputed passage has no curative value. If Fig. 14 was supposed to show that the tinctures were unnecessary, by expressing *P-modal* with a non-tinctured graph, then it failed its purpose. For Fig. 14 will be true if anything at all—a lemming, for instance—commits suicide.[17] And this is certainly an inadequate interpretation of *Ppn P*.

(3) The disputed passage contains the same line of reasoning on behalf of the tinctures that was examined above in connection with Figs. 3 to 9. I cannot find a fallacy in it. And as was pointed out in connection with Fig. 10, the need for the tinctures was not based on any one argument. Now the reference in the letter to "nearly two years after the publication" (of Prolegomena) might indicate that Peirce had in mind the very passages of the *Logic Notebook* which led us to Fig. 10. But remember that those considerations led Peirce not to abandon the tinctures, but to produce still another argument in their defense (see the final quotation in item 2, p. 101).

After repeated examination which was as careful as I could make it, and fully aware that I myself may be another over-fatigued computor, I have come to the conclusion that there is no fallacy in the disputed passage and no need for Fig. 14. I do not see that the tinctures are a mistake, but consider them a splendid tool for any philosopher who wants to be "architectonic beyond what Kant dreamed of" (Ms 280, p. 18).

6.3 SECOND GENERATION TINCTURES

The conventions and rules to be suggested in this section are put forward as provisional only, since they represent nothing more than a first attempt to

17 The reader is encouraged to construct his own proof of this. It will be very similar to the proof of Fig. 3 from Fig. 4.

extend the use of the tinctures beyond Prolegomena and Ms 670. It is not known whether they will stand the test of further research, and it seems likely that they do not cover all cases that might arise. A full scale comparison of the tinctured graphs with contemporary modal logic awaits separate treatment.

6.31 *Conventions*

At least the following changes in EG conventions (refer to Appendix 3) are called for: In C1, C2, and C3, replace 'sheet of assertion' by 'Phemic sheet'. In C7 and C8, replace 'individuals' by 'entities'; it may be that this change renders inappropriate the second occurrence of the word 'identity' in these two conventions. Replace C6 by the following: C6+. *The scribing of a heavy dot or unattached line on some part of the Phemic sheet will denote a member of whatever universe is represented by the tincture of that part of the sheet.* Thus, if a line is scribed on a metal, it will refer to an existing object of some universe of actuality; if scribed on a color or a fur, it will refer to an abstraction of whatever sort the specific tincture represents.

We must now decide upon a method for interpreting a line of identity which extends from one province (continuously tinctured surface) to another. If the different provinces are on different areas, separated by cuts, there is no problem; the line will be interpreted endoporeutically, as usual, and the province of its outermost extremity will determine the status of the entity in question. But if the line extends into different provinces on its least enclosed area, ambiguity arises. One thing we must avoid is identifying existing individuals with abstractions; so we adopt a new convention which avoids that, but which leaves other relationships to be determined later. C11. *For the interpretation of a line of identity which extends from metal to color or from metal to fur, metal takes precedence: that is, the line does not denote the abstraction (represented by the color or fur), but denotes an existing individual to whom the abstraction pertains.* Thus Fig. 1 asserts that some student of Plato possesses wisdom (following Peirce in Ms 459,

Fig. 1

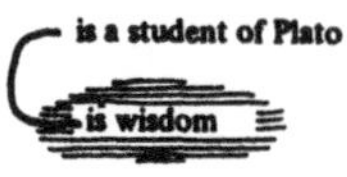

p. 34, I classify the quality wisdom as a logical possibility). Now in some cases a reference to C11 can be avoided by introducing a dyad spot and carefully piecing together the tinctured surfaces–carefully using the colored pencils. Instead of Fig. 1, for example, we may scribe Fig. 2, which also is read 'Some student of Plato has (possesses) the quality of wisdom.'

Fig. 2

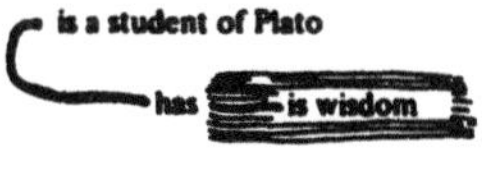

A line of identity which extends from color to fur will, I suspect, have to be interpreted case by case. To express the proposition 'The quality of mercy is not strained', I scribe Fig. 3, using argent, azure, and potent. Literally, it denies that the quality of mercy and anything that is a compul-

Fig. 3

sion pertain to each other.

In Fig. 4, sable (some shade of gray) is used on the second area to express a kind of rational necessity. The proposition is, 'People who live in glass

Fig. 4

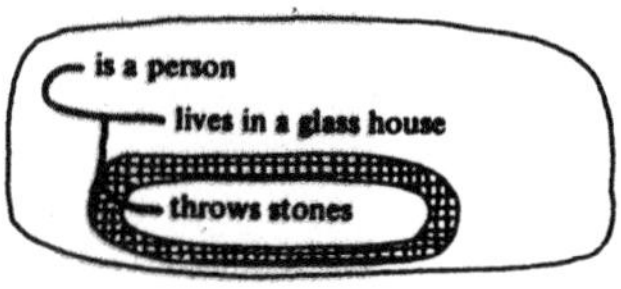

houses should not throw stones'. Although the ligature extends into three different provinces, its denotation is unambiguous because its least enclosed part is wholly contained in one province.

In Fig. 5, gules (some shade of red) is used to express an objective type of possibility. The proposition is, 'You can lead a horse to water but you cannot make him drink'. The lines of identity, because their outermost

Fig. 5

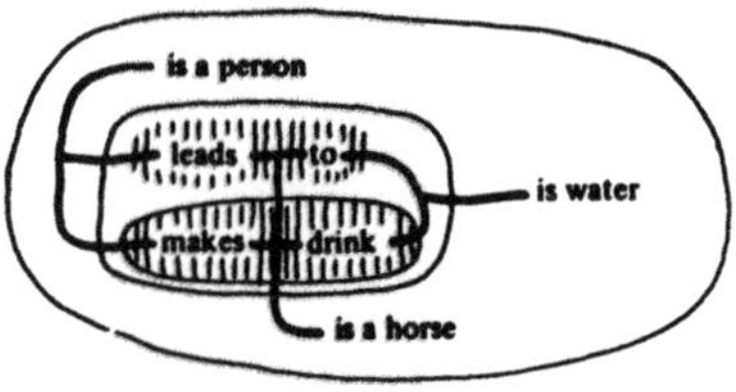

parts are wholly contained in a single province whose tincture is metal, denote individuals of the actual universe. The tinctures of color through which the lines pass do not alter this denotation, but indicate that certain possibilities pertain to the denoted individuals. The white (metal) space between the colored area and the colored enclosure is required to indicate that the modality attaches to the conjuncts and not to the conjunction 'but'.[18]

18 Here is a literal reading: Take any person x, any horse y, and any water z; it is possible for x to lead y to z, but it is not possible for x to make y drink z. Note that without the possibility tincture Fig. 5 would assert that everybody leads every horse to all the water there is (quite an accomplishment in itself), but fails to make them drink.

Fig. 6 presents an abstract statement of a familiar piece of moral suasion: 'Whoever is compelled to go one mile is commanded to go two'. Potent

Fig. 6

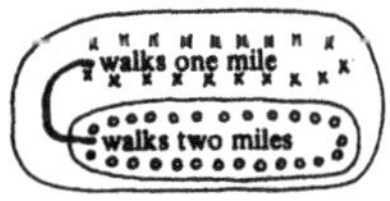

(some shade of orange) expresses what is compelled, and vair (some shade of brown) expresses what is commanded. Literally, the graph asserts that the commandment to walk two miles pertains to anyone who is under the compulsion to walk one. Note that Fig. 7, because no part of the line is on metal, diagrams a quite different proposition: 'Every compulsion to walk one mile is a commandment to walk two'.

Fig. 7

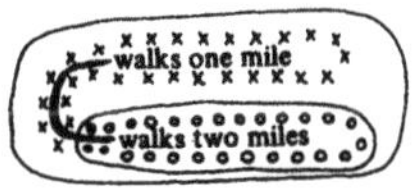

6.32 *Rules*

In this first attempt to provide rules of inference for the tinctures, I retain the rules R1-R5 with a new restriction attached to the rule of the double cut; namely, R5 will apply only to double cuts whose place and whose first and second areas are metal.[19] I add three new rules which, in their present formulation, have to do with metal and color only. The numbering continues that of earlier chapters.

R8. *Any graph, evenly enclosed on metal, may be tinctured with color.*

R9. *Any graph, oddly enclosed on color, may be tinctured with metal.*

R8 permits the transformation of Fig. 1 into Fig. 2; that is, from 'P is

Fig. 1

Fig. 2

true' it follows that P is possible. Since a blank also is a graph, this rule permits part or all of an evenly enclosed metal area to be tinctured in color. Thus, from Fig. 3, which asserts that if P is true, then the pseudo-

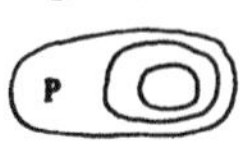

Fig. 3

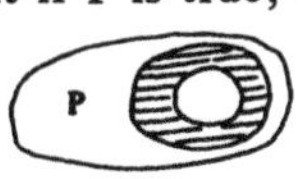

Fig. 4

19 As stated earlier, this material is put forward provisionally. Further experimentation may show that additional restrictions are necessary to preserve the consistency of the system. I fully expect that new rules will be called for. By publishing these preliminary results I hope to elicit help in the development of the tinctured EG.

graph is true, we obtain Fig. 4: 'If P is true, then the pseudograph is logically possible'.

R9 permits the transformation of Fig. 5 into Fig. 6; that is, from

Fig. 5

Fig. 6

'P is not possible' we can infer 'P is false'. Also, from Fig. 7, 'P is necessary',

Fig. 7

Fig. 8

P

Fig. 9

we can infer Fig. 8 by R9; and from this it follows that P is true (Fig. 9) by R5. Similarly, from 'It is not possible that both P and not-Q are true' (Fig. 10), we can obtain 'P materially implies Q' (Fig. 11).

Fig. 10

Fig. 11

R8 provides us with a facile way to diagram inferences by hypostatic abstraction. For example, from Fig. 12, 'A pear is ripe', we obtain Fig. 13,

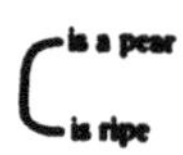

Fig. 12

Fig. 13

'A pear possesses ripeness'. And from 'Opium puts people to sleep' (Fig. 14) we infer 'Opium has dormitive virtue' (Fig. 15). Fig. 15 can also be read, 'Man is susceptible to soporifics'.

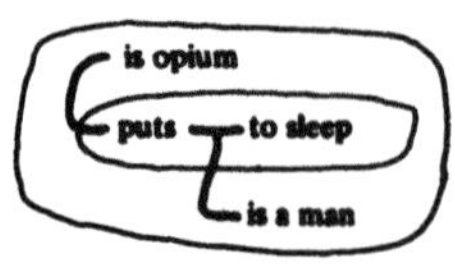

Fig. 14

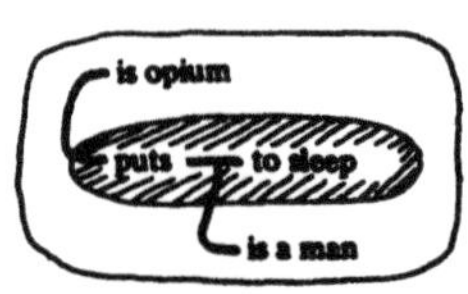

Fig. 15

There appears to be no difference in meaning between the graphs

Fig. 16

Fig. 17

of Figs. 16 and 17; both assert that P is possible. So we introduce a rule of inference which permits their transformation into each other:

R10. (a) *Any province tinctured in color can be transformed into a province of metal having a border in that color; and* (b) *a province tinctured in metal but with a border in some tincture of color can be transformed into a province of that color.* The width of the border may be adjusted to the needs of the problem in hand.

R10 enables us to give the proof which was promised in section 6.212 above, the proof of a graphical equivalence from which it will follow that Figs. 6 and 7 of that section are equivalent. We show, namely, that Figs. 18 and 19 are equivalent in the sense that each can be derived from the other.

Fig. 18

Fig. 19

Fig. 18 means 'It is not possible that both P and not-Q are true', and Fig. 19 means 'It is necessarily the case that P implies Q', or, 'P strictly implies Q'.

First we show that from Fig. 18 we can infer Fig. 19.

1. 

Fig. 18, by R10(a), or by R9.

2. 

1, by R5.

3. 

2, by R10(b).

Now, in order to obtain Fig. 18 from Fig. 19, we need only reverse the steps of the foregoing proof, justifying the passage from step 3 to step 2 by R10(a) or by R9; that from step 2 to step 1 by R5; and that from step 1 to Fig. 18 by R10(b).

7

GRAPHICAL ANALYSIS AND OUTCROPPINGS

7.1 THE PURPOSE OF EXISTENTIAL GRAPHS

In a letter to P. E. B. Jourdain, dated December 5, 1908, Peirce recalls sending to Cantor a proof that there are more collections of members of any collection than there are members of that collection. He repeats the proof to Jourdain, and then writes the following:

> I have made no other contribution to Cantor's theory, from lack of mathematical ingenuity, my forte consisting in logical analysis. I have a complete theory of this process, including its methodeutic, which I base upon my *existential graphs* which is my *chef d'œuvre* [Ms L 230a].

This is probably the source for the phrase *"My chef d'œuvre"* which occurs on p. 291 of *CP* 4.[1] Recently Professor Max H. Fisch has pointed out that the sentence in which the phrase occurs is badly constructed and perhaps insufficiently punctuated.[2] It could be taken to mean that Peirce's *chef d'œuvre* is his complete theory of logical analysis (including its methodeutic), which *he* bases on EG although it might conceivably be based on some other system of logic. Or it could be taken to mean that EG is Peirce's *chef d'oeuvre*, as the printing in *CP* suggests. The larger context of the several drafts of the letter does not decide this issue, but it *is* made clear that the connection of prime importance to Peirce is that between EG and logical analysis.

Now by a logical analysis of an inference or proposition or concept, Peirce meant a dissection (Ms 498) or a picking to pieces (4.622) of the structure of that subject matter. And when he introduced EG to his audience at the Lowell Institute in 1903, he made clear at the start that the

1 To the best of my knowledge there is no other passage in the manuscripts which could be a source for the phrase. That the letter was written to Jourdain was established by Max H. Fisch.

2 In a letter to the author, dated June 4, 1971.

purpose which the system was designed to fulfil was "to enable us to separate reasoning into its smallest steps so that each one may be examined by itself" (Ms 455, p. 2). The aim was not to facilitate reasoning, but to facilitate the study of reasoning. Moreover, this was the purpose that had previously motivated the construction of his logical algebras (Ms 499), so that his shift of interest from algebras to graphs cannot be explained as a change of program. Rather, he simply became convinced that the graphs could do a better job of analysis than the algebras could. Let us now take a look at the graphical analysis itself.

7.11 *The Graphical Analysis of Inference*

What are the smallest steps into which a reasoning can be dissected? Insertions and omissions. According to Peirce, if each elementary operation of a symbolic logic is either an insertion or an omission, then the operations of that logic are "as analytically represented as possible" (4.374). The idea of this analytic reduction of logical transformations was not new with the graphs: it was implicit in the definitions of logical composition and aggregation (3.199) that Peirce had given in 1880 (4.280), although its first "virtual enunciation" was given by O. H. Mitchell in 1883 (Ms 905, p. 38). But it was EG that seemed extraordinarily well designed to reflect this analysis of inference. The following illustration will make this clear.

The truth first enunciated by Mitchell is that the passage from premisses to conclusion "in the manner which is alone usually called necessary reasoning, can always be reached by adding to stated antecedents and subtracting from stated consequents" (Ms 905, p. 38). That is, a "logically necessary reasoning" is one in which "the substance of the conclusion can be reached from the copulate premiss" by such changes as convert 'If A is true then either C and D may be true though E is false or else F and G are true', into 'If A and X is true then either C may be true though E and Y are not both true or else F is true' (*Ibid.*). The inference from the first of these propositions to the second is valid, but this is hardly obvious. And putting it into algebraic symbols may not be much of an improvement:

1. $A \supset [(DC \sim E) \vee (FG)]$

$\therefore$ 2. $AX \supset [(C \sim (EY)) \vee F]$

But express these propositions in EG:

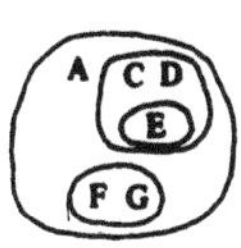

Fig. 1

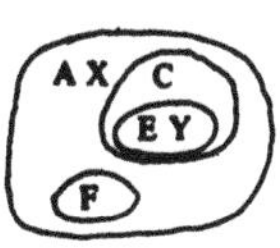

Fig. 2

Now mere inspection suffices to determine that Fig. 2 can in fact be derived from Fig. 1. For the differences between the two graphs can all be explained in terms of insertions onto odd areas (justified by R2) and erasures or omissions from even areas (by R1). The reference to antecedents and consequents should also be clear, since in EG the antecedent of a conditional is scribed on the first or oddly enclosed area of a scroll, and the consequent on the second or evenly enclosed area.

Peirce's favorite example of the analysis of inference into successive insertions and omissions is its application to the categorical syllogism in Barbara. The derivation is given in 4.3 above, and it is not difficult to see that each step is a kind of omission or insertion, justified by one of the five rules. In step 2, for example, a proposition already asserted is exhibited in a new connection by its insertion within more cuts. This is an example of experimentation on the diagram, and according to Peirce it is by observation of the results of such experimentation that new relationships can be brought to light (3.363).

I make two further observations.

(1) When Peirce in 1893 turned his attention to the psychological theory of association, he came up with something remarkably similar to the analysis of inference just presented. He accepted the usual two principles of association, contiguity and resemblance, but he split "the suggestion of B by A into two operations, one leading from A to AB and the other from AB to B" (7.393). To illustrate this, he performed a mental experiment; namely, he looked out the window, saw the cow whose milk he and his wife generally drank, and observed the following sequence of ideas:

I imagine I see a boy sitting by the cow milking her. The boy, and the stool, and the pail are added to my idea. Thence, I imagine that boy carrying the pail to the house. The cow and stool have dropped out. The straining of the milk presents itself to my imagination. A bowl is there and the pail. The boy is standing by; but I lose sight of him [7.428].

Studying that series of mental events, Peirce observed that as each new idea was added, there was always "something identical carried along" from before (7.429). The boy approaching the house with the pail was thought of as the same boy, the pail as the same pail, that he had just been thinking of. "To one skeleton-set another is added to form a compound set. Then, the first, perhaps, is dropped and the ideas which remain are viewed in a new light" (7.430). Insertions, omissions, and viewing things in a new light.

(2) Peirce claimed that EG enables one to reproduce the very "motions of reasoning" (Ms 693, p. 278), "the operation of thinking *in actu*" (4.6). In addition to insertions and omissions, he probably had in mind certain iconic features of the graphs which are discussed in section 7.24 below. But reasoning for Peirce is necessarily a conscious act, since it is something which

requires deliberate approval (2.182). Hence, by 'motions of reasoning' Peirce did *not* mean the thought-process "as it takes place in the mind" (2.27), for the thinker is not aware of it:

A man goes through a process of thought. Who shall say what the nature of that process was? He cannot; for during the process he was occupied with the object about which he was thinking, not with himself nor with his motions [2.27].

Indeed, neither physiology (2.27) nor psychology (2.184) have gained much knowledge about this process, and even if it were otherwise–if, for instance, it could be established that thinking is a continuous process, as Peirce believed it to be (2.27)–such knowledge would be "entirely irrelevant to that sort of knowledge of the nature of our reasoning" which is needed for logic (2.184). What *is* relevant to logic is the argument formulated after the fact as a summing up of the thinking-process. Here is how it works.

Having completed a process of thought, a man tries to express his conclusion in an assertion which will capture "the attitude of his thought at the cessation" of the process. He then seeks to justify his confidence in this conclusion by casting about for an assertion "which shall strike him as resembling some previous attitude of his thought" (2.27). The propositions and arguments extracted in this way constitute a kind of 'self-defence' of the original process, and Peirce maintained that it is "only the self-defence of the process that is clearly broken up into arguments" (*Ibid.*). By 'motions of reasoning' and 'operation of thinking', then, Peirce meant the elements of this self-defence.

Now all thought is dialogical and takes place in signs (4.6); hence the mind itself is a kind of sign "developing according to the laws of inference" (5.313). The Phemic sheet of EG, in relation to scribed graphs which are determinations of that sheet, represents the mind in relation to its thoughts, which are determinations of that mind. The mind as a comprehensive thought is represented by all the permissible transformations of the total graph. And any particular process of thought is represented by the graphical expression of the appropriate self-defence. Hence, "the system of existential graphs is a rough and generalized diagram of the Mind" (4.582).

7.12 *The Graphical Analysis of Propositions*

According to Peirce, EG classifies propositions into "hypotheticals (in the old and broad sense, including conditionals, copulatives, and disjunctives), categoricals, and relatives" (Ms 1147, p. 24 of "Logical graphs"). The classification is based on the number of lines of identity which are necessary to express the proposition in question: If the graph of a given proposition contains zero number of lines of identity that proposition is a non-relative (truth-functional) proposition; if its graph contains one line of identity, the proposition is a categorical proposition; and if its graph contains more than one line of identity, it is a relative proposition (Ms 481, p. 10). Although this is a three-part distinction, the major division, reflected in the usual separation of the Alpha part of EG from the Beta part, is between propositions whose form does not require the use of lines of identity and propositions whose form does require their use. This classification of propositions is not changed by the addition of a means to treat abstractions. Rather, the addition of either the different types of lines of identity (in Gamma) or the differently tinctured sheets, introduces a cross-classification.

7.121 *Subject and Predicate.* – In the traditional syllogistic, every proposition is analyzed into a single unique subject and a single unique predicate. The terms 'subject' and 'predicate' are retained in Peirce's logic of relatives, but with a broader application. A proposition, he says, consists of two parts: the predicate, "which excites something like an image or dream in the mind of its interpreter", and the subject, or "subjects, each of which serves to identify something which the predicate represents" (Ms 280, p.32; cf. 3.467). But the analysis of a proposition into these two parts can proceed in several different ways. For although any analysis of a proposition will show it to have one predicate and one only, different analyses will produce different predicates. This is illustrated in the following passage from the 1903 manuscript *Logical Tracts. No. 2.*

Let a heavy dot or dash be used in place of a noun which has been erased from a proposition. A blank form of proposition produced by such erasures as can be filled, each with a proper name, to make a proposition again, is called a *rhema*, or, relatively to the proposition of which it is conceived to be a part, the *predicate* of that proposition. The following are examples of rhemata:

—— is good
every man is the son of ——
—— loves ——
God gives —— to ——

Every proposition has one predicate and one only. But what that predicate is considered to be depends upon how we choose to analyze it. Thus, the proposition

God gives some good to every man

may be considered as having for its predicate either of the following rhemata:

___ gives ___ to ___
___ gives some good to ___
___ gives ___ to every man
God gives ___ to ___
God gives some good to ___
God gives ___ to every man
___ gives some good to every man
God gives some good to every man.

In the last case the entire proposition is considered as predicate. A rhema which has one blank is called a *monad*; a rhema of two blanks, a *dyad*; a rhema of three blanks, a *triad*; etc. A rhema with no blank is called a *medad*, and is a complete proposition. A rhema of more than two blanks is a *polyad*. A rhema of more than one blank is a *relative* [4.438].

Strictly speaking, the above examples, scribed with the line of identity, are not 'blank forms' of propositions. Hence, strictly speaking, they are not examples of rhemata. However, at the point of the exposition in which this passage occurs, Peirce had not yet introduced the line of identity, and he used this passage to begin his explanation of the line by means of what we have called Conventions 6, 7, and 8.

The rhema, or spot, corresponds to the propositional function of modern logic. Neither are propositions, both are blank forms which become propositions if proper names are put into their blanks, or if the selective pronouns 'something' or 'anything' are attached. In many contemporary logics, however, the propositional function can be expressed in the notation as a legitimate sentence of that notation (that is, as a 'well-formed formula'). But in EG a rhema is not a graph unless it is scribed with a line of identity (or a heavy dot) at each of its hooks. The upshot of this is that in EG it is not possible to express a formula with free individual variables.[3]

7.122 *Irreducibility of Triads.* – Peirce makes use of the foregoing analysis to illustrate his "remarkable theorem" that "every polyad higher than a triad can be analyzed into triads, though not every triad can be analyzed into dyads" (Ms 439, p. 16). Consider, for example, Figs. 1, 2, and 3 representing monadic, dyadic, and triadic spots, respectively. Joining two

Fig. 1

Fig. 2

Fig. 3

3 The selectives, used to abbreviate the line of identity (see Chapter 4, at the end of section 1), at first sight look like free variables, but are not; the quantification is 'implicit'. See Zeman (1967). Astonishingly, I took this 'upshot' of EG to be a defect in Roberts (1963) and (1964), in spite of the fact that I was acquainted with such systems as that in Quine (1955) in which free variables may not occur in theorems. Fortunately, J. Jay Zeman saw further; see section 7.22 below.

monadic spots, as in Fig. 4, produces a medad; it is not difficult to see that joining a monad spot to a dyad spot will produce a monad spot, since it

Fig. 4

Fig. 5

Fig. 6

leaves only one loose end; and joining a monad to a triad will produce a dyad. Joining two dyad spots will produce another dyad, as in Fig. 5; and joining two loose ends of the same triad will produce a monad, as in Fig. 6. If two triads are joined as in Fig. 7, a dyad results; but if they are joined as

Fig. 7

Fig. 8

in Fig. 8, a spot with four hooks results. Adding additional triads will produce spots of any adicity.

It might be thought that two dyads can be joined in such a way as to produce triads, as in Fig. 9, for instance. But this depends upon the existence at the arrow of a point of teridentity (a triad), without which there

Fig. 9

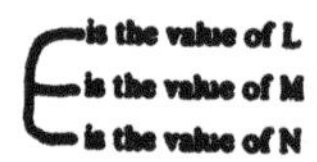

Fig. 10

could be no such join. Consider also the inference 'L and N are co-equal to M, hence, L and N are equal to each other'. Algebraists will represent this in the two equations L = M and M = N, from which L = N is concluded by virtue of Euclid's first axiom. But in EG the premiss is represented as in Fig. 10, where again a point of teridentity is required (Ms 300, p. 33). And according to Peirce, such a point *is* required, since "teridentity is not mere identity. It is identity *and* identity, but this 'and' is a distinct concept, and is precisely that of teridentity" (4.561). Incidentally, the conclusion is inferred from Fig. 10 by R1 (erasure of the spot " ——is the value of M"; cf. 1.346).

7.123 *Composition of Concepts.* – Another consequence of Peirce's analysis of propositions is a solution to the "puzzle" of how concepts are combined.

> Suppose two concepts, A and B, to be combined. What unites them? There must be some cement; and this must itself be a concept C. So then, the compound concept is not AB but ACB. Hereupon, obviously arises the question how C is combined with A or with B. The difficulty is obvious, and one might well be tempted to suspect that compound concepts were impossible, if we had not the most manifest evidence of their existence [Ms 498, pp. 28-29; cf. Ms 499].

Now Peirce assumes that all propositions can be *analytically* expressed in EG (4.583), that is, that the parts of graphs are related to each other "in forms of relation analogous to those of the assertions they represent" (Ms L231, p. 15; microfilmed with Ms 514). Hence he concludes that what EG represents to be true of propositions and concepts must be true of them (4.583).

What EG tells us about the composition of concepts is this: "each component must be indeterminate in some respect or another; and in their composition each determines the other" (4.572). The components "supply each the other's lack" (*Ibid.*). Here there is no attempt to explain one instance of a combination by replacing it with two such instances. There is, indeed, no cement which unites two (or more) concepts together. Rather, they fit together like the pieces of a jig-saw puzzle; or they mesh together, like teeth on gears. The fit is accomplished by "the minimum possible number of modes of logical combination, –namely, one only, that in which a hook of a graph is joined to a single other hook of a graph".[4] That is, to render any spot determinate, each of its hooks must be attached to a hook of another graph. Yet no concept becomes so determinate that it cannot be made more determinate (4.583), since it is always possible to increase our knowledge. And this is reflected in R3 which permits a branch with a loose end to be added to any line of identity, at any point on that line.[5] The general doctrine, that things indeterminate in themselves function to determine each other, applies also to propositions, which partially define each other on the recto and partially limit each other on the verso. 'Partially', because, like concepts, propositions cannot be "perfectly determinate" (4.583).

According to this explanation concepts are combined in very much the same way that chemical elements are pictured to combine in the doctrine of valency. It was shown in Chapter 2 that Peirce's logic diagrams were consciously modeled after the diagrams used by chemists to represent the composition of elements. According to the doctrine of valency current when Peirce learned his chemistry, every element has a fixed number of loose ends (3.470) or unsaturated bonds (3.469), and a compound is formed when the loose ends of one atom are joined to the loose ends of other atoms. In this way each atom may be said to partially determine the other. Changes in the theory of valency since Peirce's time do not disturb the basic analogy; whether chemical composition is regarded as the transfer of electrons from one atom to another or as the sharing of electrons between

4 Ms 296, p. 8. Hooks are joined, of course, by bringing their lines of identity "into abuttal". Ms L 463, a draft of a March 1906 letter to Lady Welby, p. 37.

5 On p. 34 of the letter draft referred to in footnote 4, Peirce remarks that "every line of identity ought to be considered as bristling with microscopic points of teridentity; so that —— when magnified shall be seen to be 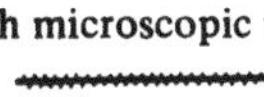".

atoms (co-valency), the atoms may still be said to partially determine each other. And even if it drops out of chemistry altogether, the notion of valency may still have philosophical application.

7.2 OUTCROPPINGS

The rest of this chapter is devoted to some relatively unconnected features of EG, plus concluding remarks.

7.21 *The Graphical Notation*

EG has no symbol whose sole purpose is to indicate the grouping of expressions. In this respect it is similar to the bracket-free notations such as the one invented by Jan Łukasiewicz in 1929 (Łukasiewicz [1957], 78). One major advantage of such notations, according to A. N. Prior, is economy: "no special rules about bracketing and rebracketing need to be included among the rules for proving one formula from another" (Prior [1962], 6).

Even when no special symbol is required, however, grouping is accomplished in some fashion in all these systems. In EG, the cut does the job; but it also expresses negation and indicates quantification. Note, however, that although the cut separates graphs on one area from those on another area, it does not impose any ordering onto graphs scribed on the same area. No symbol or convention does this for EG: "Operations of commutation, like $xy \therefore yx$ [and, as Peirce goes on to show, operations of association, like $(xy)z \therefore x(yz)$], may be dispensed with by not recognizing any order of arrangement as significant" (4.374).

These simplifications have two important results. In the first place, the fact that EG requires only two special symbols (the cut and the line of identity) plus the sheet of assertion accounts in part for the unusual ease with which inferences can be drawn in Alpha and Beta. Peirce did not design the graphs for this purpose, but in spite of that the graphical operations—adding or removing the double cut, writing any proposition on any odd area, erasing any proposition from any even area, extending the line of identity inward through cuts—can all be performed quickly and easily. Whitehead put the point nicely: "By relieving the brain of all unnecessary work, a good notation sets it free to concentrate on more advanced problems, and in effect increases the mental power of the race" (Whitehead [1948], 39).

In the second place, a graph of even meager complexity can be read in several different English sentences, if only the reader will keep in mind two or three basic patterns. Fig. 1 is the pattern for an alternative proposition, and Fig. 2 is the pattern for a conditional. Just imagine what Fig. 3 would look like with a double cut surrounding P, and it will be obvious that 'Either not-P or Q' means the same as 'If P then Q'. A student of EG will learn automatically many of the linguistic equivalences that require an excess of time and symbols in algebraic notations.

Fig. 1

Fig. 2

Fig. 3

7.22 *EG Axioms and Natural Deduction*

At the turn of the century, in a draft of a letter to his former student Christine Ladd-Franklin, Peirce wrote: "You ask whether Logical Graphs have any bearing on Non-Relative logic. Not *much*, *except* in one highly important particular, that they supply an entirely new system of fundamental assumptions to logical algebra" (Ms L 237). It is safe to assume that Peirce was thinking of the Alpha part of EG. And since the first completeness proof for the propositional calculus was not discovered until 1921 (Emil Post [1921]) Peirce presumably had in mind only that Alpha was as complete as other formulations of non-relative logic. It is nevertheless true that Alpha is a complete propositional calculus, and a rather economical one too, since the rules of Chapter 3 together with SA as a single axiom (see the discussion leading up to C1) are all that is required. A proof of this is given in Appendix 4.

When the unattached line of identity (or heavy dot) is added as a second axiom–and this was given in C6 of Chapter 4–Beta turns out to be a complete functional calculus of the first order (with identity), a remarkable fact discovered by J. Jay Zeman (Zeman [1964]). A proof of this is given in Appendix 4, also.

According to C6, the unattached line merely asserts that some individual object exists. The same assumption is made in *Principia Mathematica* and is expressed there in theorem *10·25: $(x)Fx \supset (\exists x)Fx$. Russell and Whitehead translate this as "what is always true is sometimes true", and they add the following commentary:

> This would not be the case if nothing existed; thus our assumptions contain the assumption that there is something. This is involved in the principle that what holds of all, holds of any; for this would not be true if there were no "any".[6]

This *Principia* theorem has a straightforward translation in EG (Fig. 1), but finding an algebraic equivalent of the Beta axiom is slightly complicated by

6 Whitehead and Russell (1927), I, 20. 'Presuppositionless' logics which do not make this assumption have been investigated in Hintikka (1959), Leblanc and Hailperin (1959), and Leonard (1956).

Peirce's reading of it as "something exists"(Ms 455, p. 21). The usual practice among logicians is to bring in predicates of some sort when existence is

Fig. 1

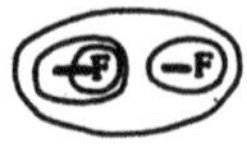

asserted of anything: "When, in ordinary language or in philosophy, something is said to 'exist', it is always something *described*" (Whitehead and Russell [1927], I, 174). Peirce recognized that his axiom was a bit peculiar, for he remarked that "it would be a nice question" whether it has "any positive meaning or not" (Ms 454, p. 15); but he always thought of it as a graph, and therefore as an assertion of some sort. In any case, since the line asserts the numerical identity of the individuals denoted by its extremities (C7), the formula $(\exists x)[x = x]$ suggests itself as an algebraic translation of the Beta axiom. This can be read 'something is self-identical'.

In spite of the fact that SA and the unenclosed, unattached line of identity—*and* the double cut[7]—are correctly viewed as axioms, Peirce did not develop EG as an axiomatic calculus in the style of *Principia*, of Church (1956a), of Quine (1955). His most formal presentations of the system result in additional rules of inference (metatheorems), not additional permissible graphs (theorems).[8] And the method of proof throughout these presentations is that of "proving an implication by making an assumption and drawing a conclusion"—which is precisely the way Alonzo Church characterizes the method of natural deduction (Church [1956a], 165).

Natural deduction systems were first introduced into the literature in 1934 by Gerhard Gentzen and Stanisław Jaśkowski.[9] Such systems ordinarily make no use of axioms, but employ the 'rule of conditionalization' (the deduction theorem) as a primitive rule of inference. It is this use of the rule that is credited to Gentzen and Jaśkowski (Church [1956a], 164); the rule itself was known to the Stoics (Kneale and Kneale [1962], 170). Peirce was familiar with the rule. He had used it (informally) to establish implications by deriving consequences from assumptions,[10] and in

7 4.415, 567. The double cut as "a blank proposition always true" is also listed as an axiom of an algebraic system which was modeled after EG in *LN*, p. 179r (December 11, 1900).

8 See *LN* pp. 114r-124r (June 1898); Ms 484 (August 1898); 4.485-498 (1903); Ms 478, pp. 165-168 (1903).

9 Gentzen (1934-1935), and Jaśkowski (1934). Jaśkowski points out (on p. 5) that his first results were obtained eight years earlier when he was a student of Łukasiewicz.

10 E.g., when discussing the third icon in his 1885 article "On the Algebra of Logic", Peirce reasons that since $x \prec z$ follows from $(y \prec x) \prec z$, we may write $[(y \prec x) \prec z] \prec (x \prec z)$, 3.380. Cf. 4.485-498, which is part of *Logical Tracts. No. 2*, written in 1903.

1898 he proved it as a metatheorem of EG.[11] Two years later he thought of using this rule as a primitive rule of inference. The idea occurred to him while he was experimenting with various statements of basic rules for his graphs and for algebraic systems loosely modeled after EG. On page 180r of his *Logic Notebook* (dated December 11, 1900), Peirce presented a list of primitive rules of inference which included the following statements of the rule of conditionalization (the claw symbol, —<, represents material implication):

(a) If A written alone is transformable into B, we can write alone A —< B.
(b) If A —< B is scriptible alone A written alone can be transformed into B.

It is clear that Peirce here anticipated Gentzen and Jaśkowski in this one particular, namely, with regard to the notion of using the rule of conditionalization as a primitive rule of inference. However, he does not seem to have done much with it, and the expositions of EG which became standard do not contain this rule as a primitive rule of inference.

Nevertheless, when Alpha is structured as a natural deduction system, the result is a deductively complete propositional calculus. To show this, we first delete C1 from our list of conventions, since it is by C1 that SA is a graph; and if SA is a graph, it functions as an axiom, and completeness follows as in Appendix 4 without need of the rule of conditionalization. So we begin with Alpha, minus C1, but augmented by rule (a) above, stated as follows:

(a) If P can be transformed into Q, we can scribe  on SA.

To show that this version of Alpha is complete, it is sufficient to obtain the double cut as a theorem; for once this is done the completeness proof in Appendix 4 applies. The first three steps are bracketed on the left to indicate that they constitute the subordinate proof of the antecedent of rule (a). Strictly speaking, the first graph that we are entitled to write on SA is that in step 4.

1.	P	Assumption.
2.		1, by R5.
3.	P	2, by R5.
4.		1-3 by (a).

11 The proof occurs on p. 118r of the *Logic Notebook*. The rule is stated as follows: "If one graph can be illatively transformed into another an enclosure may be written consisting of an oval enclosing the former graph and an oval enclosing nothing but the latter".

5.  4, by R5.

6.  5, by R4, deiterating the twice enclosed P.

7.  6, by R1.

8.  7, by R4.

7.23 *The Categorical Syllogism*

EG provides a particularly neat treatment of the categorical syllogism, without the assumption of existential import. For (1) the operations of conversion, obversion, and contraposition involve nothing more than the addition or removal of double cuts, and (2) the validity of a syllogism can be determined by mere inspection.

(1) The first row in Fig. 1 below gives the graph of the four categorical propositions (properly labeled); the second row gives the obverse, the third the converse (where applicable), and the fourth the contrapositive (where applicable) of the propositions diagramed in that first row.

Fig. 1

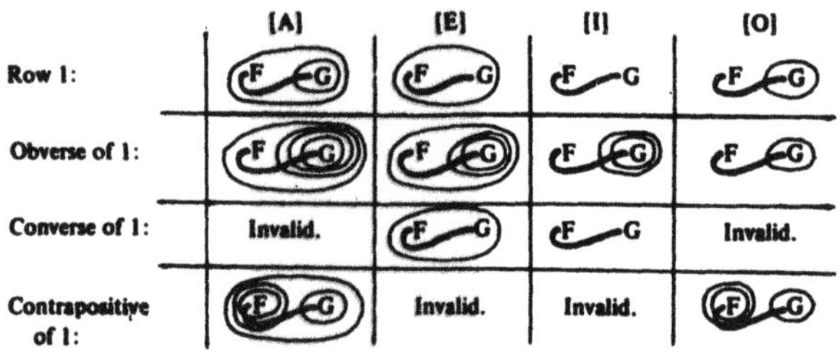

The only 'trick' to translating these graphs into English, is in recognizing that a graph of the form displayed in Fig. 2 can be read in two ways: 'There

Fig. 2

is something which is not G', and 'There is something which is a non-G'. This explains why the same graph can express both the O proposition and its obverse. To understand the converses of the E and I propositions, it is sufficient to recall that the order or arrangement of graphs on the same area has no significance in EG. Note how obvious it is that propositions A and O, and propositions E and I, are contradictories.

(2) One standard way to determine the validity or invalidity of a categorical syllogism is to check it against rules specially designed to reveal these properties. Such rules can be found in almost any introductory logic text, and it is an easy matter to adapt them to the graphical notation. For example:

(*a*) The middle term (the rhema which occurs in both premisses) must be once oddly and once evenly enclosed.
(*b*) A term may be evenly (oddly) enclosed in the conclusion if and only if it is evenly (oddly) enclosed in a premiss.
(*c*) No categorical syllogism is valid whose premisses each contain an odd number of cuts.
(*d*) If one premiss contains an odd number of cuts, then the conclusion must contain an odd number of cuts.
(*e*) No categorical syllogism is valid in which the lines of identity of both premisses are evenly enclosed.

The application of these rules to syllogisms expressed in graphs is frequently more immediate than the application of comparable rules to the same syllogisms expressed in other notations. Consider this argument: 'All non-Confucians are either indifferent to or contemptuous of the *I Ching*, but some philosophers are Confucian. Hence, some philosophers respect the *I Ching*'. After putting this into categorical form in traditional notations, it is still necessary to perform three operations on the first premiss in order to apply the rules (obversion, conversion, and again obversion, are used to reduce the five terms to three). But express the same syllogism in graphs, and inspection shows immediately that the middle term (the rhema 'is a Confucian') is evenly enclosed in both premisses (Figs. 3 and 4), and that the rhema 'res-

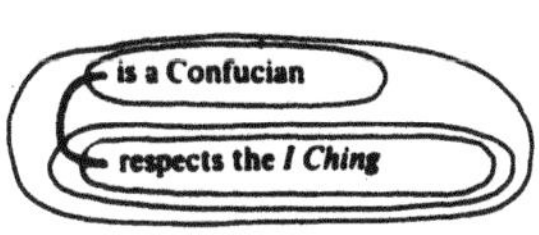

Fig. 3

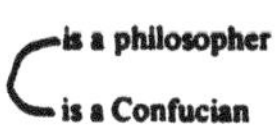

Fig. 4

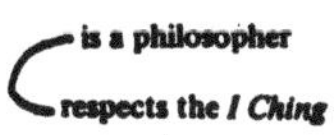

Fig. 5

pects the *I Ching*' is oddly enclosed in a premiss (Fig. 3) but evenly enclosed in the conclusion (Fig. 5). The syllogism violates both rules (*a*) and (*b*), and is therefore seen to be invalid.

To enable EG to take care of the so-called 'weakened' moods of the syllogism, those namely in which a particular conclusion is derived from two universal premisses, it is necessary only to add as an additional premiss the graph expressing the existence of objects denoted by the subject term of the conclusion. This addition is necessary in spite of the Beta axiom, for it is not enough that something exists; a syllogism requires that it be a certain something.

7.24 *Iconicity*

According to Peirce, a good system of diagrams should be iconic: the parts of the diagrams should be related to each other in the same way that the objects represented by those parts are themselves related to each other.[12] The utility of diagrammatic reasoning is its capacity to reveal unexpected truth, and this capability is a peculiarity of icons: "A great distinguishing property of the icon is that by the direct observation of it other truths concerning its object can be discovered than those which suffice to determine its construction" (2.279; cf. Ms 650).

Peirce compared logic diagrams with maps: both convey somewhat difficult and complicated matters with a clarity and ease that can hardly be matched in any other way, and both are particularly well suited for experimentation (one can rearrange and recombine the parts of a logic diagram in much the same way that one can use pins and rearrange them on a map to represent the deployment of forces in a battle) (4.530). However, just as we do not expect a map to include a marking of the position of every tree or rock in the area it represents,[13] so we should not expect a logic diagram to be an exact replica of the objects it represents. The essential thing is that the diagram shall be analogous to what it represents, "differing from the objects themselves in being stripped of accidents" (Ms 524). That is, the diagram should be constructed in such a way as to have all the characteristics of the objects represented which have any bearing on the reasoning involved, and no other characteristics (4.233; cf. 3.92).

Peirce was convinced that EG enjoyed to an unusually high degree the essential features of diagrammatic thinking. The system, he claimed,

> is truly *diagrammatic*, that is to say that its parts are really related to one another in forms of relation analogous to those of the assertions they represent, and that consequently in studying this syntax we may be assured that we are studying the real relations of the parts of the assertions and reasonings; which is by no means the case with the syntax of speech [Ms L 231, p. 15].

I suspect that Peirce had in mind not only the graphical analysis of inference and propositions already described (7.1 above), but also certain other ways in which the graphs may be said to be iconic.

Consider the Phemic sheet, which usually functions as a sheet of assertion, SA. It is a two-dimensional continuity (see 4.561n.1). Peirce tells us that the sheet may be imagined to represent the continuum of all true propositions. It seems necessary to think of this totality as a continuum, "since facts blend into one another" (4.512). Furthermore, we sometimes deal with universes other than "the universe of existent individuals"—namely, all

12 3.363 (1885); 3.418 (1892); 3.556 (1898); 3.641 (1901); *LN*, p. 286r (September 5, 1906).

13 Professor Fisch recalls working with Chinese maps of small areas which not only showed every rock and tree but attached a short poem to each.

those represented by the tinctures–so that the continuum of truths is to be conceived as having "more dimensions than a surface or even than a solid" (4.512; cf. 514). The two-dimensional tinctured surfaces provide a much-simplified representation of this continuum, but still, in EG, we use one continuum to represent another.

When Peirce said in 1898 that his first system of graphs, the entitative system, was "quite unnatural in some of its leading conventions" (Ms 484, § 2), he was referring to such things as the interpretation of juxtaposition as alternation (see 2.4 above). EG, for which juxtaposition is conjunction, is more iconic; to scribe two graphs is like asserting them both (4.434). In fact, to write down two propositions is to assert them both, if the writing is submitted to a notary or a court of law or a university examiner. The conditional proposition 'If P then Q' asserts neither P nor Q; in EG the scroll provides a means for scribing the conditional without placing either P or Q on the Phemic sheet itself. The cut provides an iconic way to diagram false propositions, for it fences off its contents from the surface of the sheet.

The line of identity asserts the numerical identity of the individuals denoted by its extremities, and the continuity of the line is an excellent likeness of the identity thus asserted (4.448, 561n.1). In Quine (1955), 70, use is made of similar lines to illustrate the pronomial reference of letters (such as x and y) in algebraic formulations of quantification. The proposition 'No man has seen every city' is expressed semi-algebraically as follows:

$$(x)\,(\,x \text{ is a man} \supset \sim (y)\,(y \text{ is a city} \supset x \text{ has seen } y))$$

In this formula the letters x and y serve "merely to indicate cross-references to various positions of quantification". To display this Quine presents the diagram of Fig. 1.

Fig. 1

Fig. 2 is a Quine diagram of the proposition 'Whatever number you may select, it will turn out, whatever number you may next select, that the latter is less than, equal to, or greater than the former'.

Fig. 2

These diagrams serve a useful illustrative purpose, even if, as Quine suspects, they are "too cumbersome to recommend themselves as a practical notation" (Quine [1955], 70). But let us translate these propositions into existential graphs, Fig. 1 into Fig. 3, and Fig. 2 into Fig. 4.

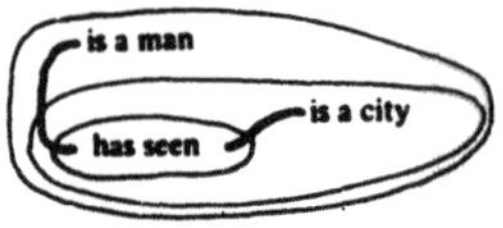

Fig. 3

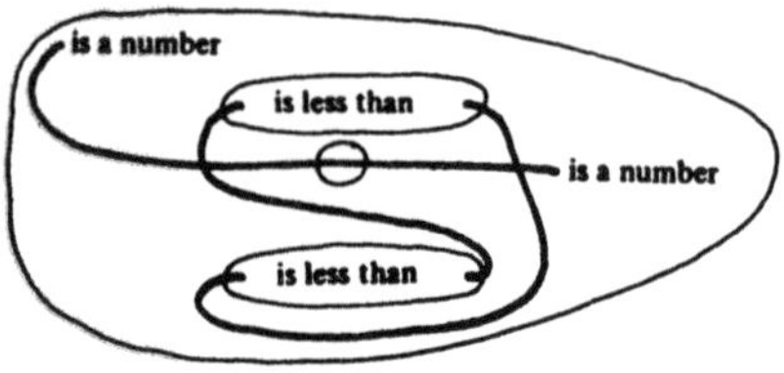

Fig. 4

Peirce's diagrams are at least as iconic as Quine's, and a good deal simpler. Figs. 3 and 4 involve only three kinds of symbol: the heavy line, the spots, and the cut. Figs. 1 and 2 involve six kinds of symbol: the line, the spots, parentheses, the horseshoe, the tilde, and the wedge (plus three signs that could as easily have been expressed in English).

Logic diagrams these days are certainly less familiar than algebraic notations of logic. But are they really too cumbersome? Would it be fair to say that they are "less perspicuous than the usual" notations, as some say of the Polish notation (Church [1956a], 38 n. 91)? Well, if a facile and perspicuous notation is one that can be quickly learned and easily manipulated, then years of experience with university students have convinced me that EG is the most perspicuous, and *Principia* notation the least. The unusual ease with which inferences can be drawn in EG is something of an unexpected bonus.

In addition to what has already been mentioned, the graphs were iconic for Peirce in a special way: "My 'Existential Graphs' have a remarkable likeness to my thoughts about any topic of philosophy" (Ms 620, p. 9). By this he certainly meant to express his preference for thinking in diagrams:

> I do not think I ever *reflect* in words: I employ visual diagrams, firstly because this way of thinking is my natural language of self-communion, and secondly, because I am convinced that it is the best system for the purpose [Ms 620, p. 8].

But I think he also had in mind that the structure of EG captures so many important features of his own philosophy. The 'likeness' was built in.

7.25 *Postscript*

In the spring of 1903 Peirce delivered a series of lectures on pragmatism at Harvard University. An early draft of the second lecture contains the following remark:

But I must tell you that all that you can find in print of my work on logic are simply scattered outcroppings here and there of a rich vein which remains unpublished.[14]

This is a happy metaphor indeed. Wide ranging and powerful, Peirce's publications suggested an extensive layer of thought still not brought to the 'surface' of public attention and containing material which would reveal the relationships between the parts of his work already visible. He supposed that most of his material had been written down, but thought that "an entirely new presentation" was required. This would have been a monumental task, and necessary support was not forthcoming.[15] As a result, although Peirce kept at it, much of the mining of that rich vein was left to others.

Now I believe that EG could play a significant role in the systematic study of Peirce's logic. In the first place, he packed into the system a great deal of his thinking about logic, and he used it to extend the reach of his logic into modalities. In the second place, he applied EG to "the most difficult problems of logical theory" (4.571). In this connection, its "true utility" is the support it gives to the mind "by furnishing concrete diagrams upon which to experiment".

Peirce also found his graphs important in relating his work in logic to his work in philosophy. He thought, for instance, that they would provide a "less mechanical and fatiguing" introduction to the logic of relatives, an introduction which would "better display its living ideas and connections with philosophy and with life" (Ms 436, pp. 28-29). Peirce included an exposition of EG in his series of articles on pragmaticism, because with the graphs his proof of pragmaticism "could best be rendered plain" (Ms 300, p. 15). In fact, all the elements of his philosophy were to come into focus together in this proof: the synechism (5.415), the realism and the theory of modality (5.453ff), the theory of signs and the distinction between deduction, induction, and retroduction (Ms 330), and more.

In view of these considerations, and the many more developed throughout this book, it should not be puzzling that Peirce placed a high valuation on his graphs.[16] The major reason for this assessment was made clear from start to finish since Peirce repeatedly stated that his purpose in constructing

14 Ms 302, p. 10. If more than this is to be quoted, then the entire passage should be quoted as in Murphey (1961), 1 n.1, unlike *CP* 2, p. 2.

15 In 1902 Peirce applied to the Carnegie Foundation (Ms L 75) for support in making such a new and comprehensive presentation of his logic. The application was rejected.

16 According to Zeman (1968), 150, Peirce attached to EG "that puzzling 'subtitle' "–*chef d'œuvre*–because he wanted to develop a logic of continua and he hoped the graphs would make this possible. But it is highly improbable that Peirce's tremendous amount of work on logic diagrams was motivated by any single application or use he hoped to make of them.

EG was to build an engine of analysis. As he developed the graphs and applied them to various problems, it was always their experimental possibilities and analytic power that chiefly pleased him. The letter to Jourdain which opened this chapter surely counts as evidence for this. As to the phrase '*chef d'oeuvre*', –whether Peirce was saying that EG was his chief work, or was the chief part of his chief work, either claim places the graphs at the center of his philosophy.

APPENDIX 1

A SELECTIVE CHRONOLOGY OF PEIRCE'S WORK ON LOGIC

The best published accounts of Peirce's work on formal logic are given by Church (1956a), Bochenski (1961), and Kneale and Kneale (1962). Lewis (1960), although written in 1918, contains a still valuable account of the historical development of symbolic logic, of which 28 pages are devoted to Peirce; and Berry (1952), though restricted in scope, is also useful for our purpose. An examination of these five works indicates that the bulk of Peirce's contributions to logic are located in fourteen papers, ranging in date from 1867 to 1902. The fourteen papers are marked with an asterisk in the following list. Apart from these, the list is limited to works which are especially relevant to logic graphs. Numbers such as G-1867-la are Burks's bibliography numbers, as given in *CP* 8.

*G-1867-la. "On an improvement in Boole's Calculus of Logic", 3.1-19. Besides suggesting certain improvements for Boole's calculus of probability (see Lewis [1960], 100ff), this article gives a clear statement of non-exclusive alternation. Peirce points out that the use of this type of alternation produces an exact parallelism between the theorems involving logical addition and those involving logical multiplication (Bochenski [1961], 303; Kneale and Kneale [1962], 422-423).

*G-1867-1d. "Upon the Logic of Mathematics", 3.20-44. Here Peirce begins to use the upper case sigma (Σ) to designate the logical (not the arithmetic) sum, and this may be an early anticipation of his later use of this symbol as the existential quantifier. In a later paper, G-1870-1, the upper case pi (Π) is used to designate the logical product; it will later be used as the universal quantifier (Berry [1952], 161).

These 1867 articles were written as part of a series on logic that Peirce published in the *Proceedings of the American Academy of Arts and Sciences*. In both, Peirce indicates how mathematical relations, operations, and systems may be derived from symbolic logic, and in doing this he anticipates (in a fragmentary, but nevertheless noteworthy manner) the approach of Whitehead and Russell (1927) (Lewis [1960], 100ff).

*G-1868-2c. "Grounds of Validity of the Laws of Logic: Further Consequences of Four Incapacities", 5.318-357. Partly because of this paper (see especially 5.340 and n.l) Peirce is credited with being one of the few logicians of that day to have a scholarly knowledge of and interest in the semantical paradoxes (Bochenski [1961], 387).

G-1869-2. A series of lectures on "British Logicians". 1.28-34 is from Lecture 1, "Early Nominalism and Realism". Peirce remarks that the successes of scientists and mathematicians are due in part to their use of observation, and he quotes Gauss's statement "algebra is a science of the eye" (1.34).

*G-1870-1. "Description of a Notation for the Logic of Relatives, resulting from an Amplification of the Conceptions of Boole's Calculus of Logic", 3.45-149. This is the only paper on our list that dates from the 1870's. It is cited for two reasons (apart from the previously mentioned use of π): (1) Peirce here works out systematically (the first such systematic treatment) the notion of class inclusion, and he introduces the claw sign $-\!\!<$ to stand for this relation;[1] and (2) the treatment given to relative terms amounts to an extension of Boole's logical algebra beyond the realm of absolute terms (beyond Aristotle's syllogistic, for example), an extension which would result in a development of the logic of relations as the logic of classes had been developed (Kneale and Kneale [1962], 428-429; Lewis [1960], 83). A calculus such as the one described in this paper would, according to Peirce, be "practically useful in some difficult cases, and particularly in the *investigation* of logic" (3.45).

1870. Benjamin Peirce defines mathematics as the science which draws necessary conclusions (B. Peirce [1881], 97). Charles remarks upon his father's definition in several places (see for instance 4.229), and in one undated manuscript he shows how it serves to distinguish the mathematician from the logician: "The mathematician's interest in reasoning is as a means of solving problems .... The logician, on the other hand, is interested in picking a method to pieces and in finding what its essential ingredients are" (Ms 78).

1878. At about this time William K. Clifford and James J. Sylvester began to use chemical diagrams to represent algebraic invariants (Murphey [1961], pp. 196-197).

1879. Peirce becomes associated with Sylvester at the Johns Hopkins University.

*G-1880-8. "On the Algebra of Logic", 3.154-251. This is the first of five papers on our list that date from the 1880's. In this one (see especially 3.204ff) Peirce gives the process of reduction to both full disjunctive normal

1 Berry (1952), 154; Bochenski (1961), 304-305; Lewis (1960), 83. As noted in section 7.22, Peirce also used the claw symbol as a sign of material implication.

form and its dual, the full conjunctive normal form (Church [1956a], 166 and n.299; see also Lewis [1960], 96 and Berry [1952], 154).

*G-c.1880-1. On a Boolean algebra with one constant (title not supplied by Peirce), 4.12-20. This paper was not published during Peirce's lifetime. In it Peirce demonstrates the possibility of developing the entire logic of propositions in terms of the single operator now called 'joint denial': "Neither P nor Q is true" (Berry [1952], 155; Bochenski [1961], 344-345; Church [1956a], 133n.207; Kneale and Kneale [1962], 423, 526). The first publication of the fact that such a possibility exists was by Sheffer in 1913—thirty-three years later (Sheffer [1913]).

*G-1881-7. "On the Logic of Number", 3.252-288. This article includes, among other things, some of the essentials of number theory (for example, informal statements of recursion equations for addition and multiplication) and a definition of simple order (Church [1956a], 322n.526 and 337n.550).

1882. Peirce to O. H. Mitchell, a letter dated December 21, 1882, Ms L 294. "The notation of the logic of relatives can be somewhat simplified by spreading the formulae over two dimensions". First notice by Peirce of a system of logic diagrams to treat of the logic of relatives.

*G-1883-7d. "The Logic of Relatives", 3.328-358. This appeared as "Note B" in the volume edited by Peirce entitled *Studies in Logic, By Members of the Johns Hopkins University* (Peirce [1883]). It is considered to be one of Peirce's clearest statements of his theory of the logic of relatives (Bochenski [1961], 377-379; Lewis [1960], 85). Here and in the next article listed Peirce "created a symbolism adequate for the whole of logic and identical in syntax with the systems now in use" (Kneale and Kneale [1962], 431). One of the major features of his symbolism is the explicit use of Σ and π as quantifiers and the use of subscripts to identify terms (Berry [1952], 161; Kneale and Kneale [1962], 430-431).

*G-1885-3. "On the Algebra of Logic: A Contribution to the Philosophy of Notation", 3.359-403. This is the only one of the fourteen papers to be mentioned by all five of our 'authorities' (first paragraph, this appendix). Of the many items of value to be found here, the following eight are singled out for attention. (1) In this paper we find the first explicit use of two truth-values and the first statement of the truth-table decision procedure as a general procedure (Church [1956a], 25n.67 and 162; Berry [1952], 158; Bochenski [1961], 329-330; Kneale and Kneale [1962], 420). (2) Peirce presents an especially clear and adequate statement of quantification theory (Berry [1952], 165; Bochenski [1961], 348-349; Kneale and Kneale [1962], 432f; Lewis [1960], 96ff), which includes his special notational contribution, namely, the use of an operator variable in connection with the quantifier (discovered independently by Peirce, but six years after Frege published it in his *Begriffsschrift*) (Church [1956a], 288n.453). (3) The article contains probably the earliest use and description of the method of transforming quantified statements which is known as reduction to prenex normal form[2] (3.396). (4) Peirce gives the modern definition of identity (3.398), avoiding Leibniz's confusion of use and mention.[3] (5) The so-called 'Peirce-Dedekind' definition of an infinite class (as one having a one-to-one correspondence with a proper subclass) is presented (Church [1956a], 344n.565; Kneale and Kneale [1962], 440). (6) Peirce suggests that the negation of a proposition p be expressed by means of the formula p $\prec \alpha$, 'If p then α', where α is said to be an index of no matter what token (3.384-385). This is an anticipation of an expression Russell was to use in 1903 and 1906 (Church [1956a], 151n.226). (7) In this same paper is what Bochenski considers to be one of the best justifications of the use of material implication in formal logic (Bochenski [1961], 313-314). (8) All deductive reasoning involves observation, and the observation usually takes the form of observation of a diagram (3.363).

1886. A. B. Kempe publishes his "Memoir on the Theory of Mathematical Form" (Kempe [1886]).

1887. Peirce to Kempe, a letter dated January 17, 1887 (the date is given in Kempe [1897], 453). This letter of criticism led Kempe to "reconsider certain paragraphs" of his Memoir and make the "amendments" of Kempe (1887).

1889. "Notes on Kempe's Paper on Mathematical Forms", Ms 714. In this unpublished manuscript, dated January 15, 1889, Peirce begins to develop a graphical system different from Kempe's, and with similarities to EG.

*G-1892-1b. "The Critic of Arguments", 3.404-424. (1) Peirce extends the concept of function to many-place functions in 3.420-421 (Bochenski [1961], 323-324). (2) Peirce again speaks of algebra as a kind of diagram (3.419), and he remarks that "unpublished studies have shown me that a far more powerful method of diagrammatisation than algebra is possible, being an extension at once of algebra and of Clifford's method of graphs; but I am not in a situation to draw up a statement of my researches" (3.418).

G-1891-1f. "Reply to the Necessitarians. Rejoinder to Dr. Carus", 6.588-618. This was published in July of 1893. Deduction is a matter of the perception of and experimentation with certain imaginary objects, such as diagrams (6.595).

2 Church (1956a), 292n.469; Lewis (1960), 98; Kneale and Kneale (1962), 432. In 3.505 Peirce says "The student ought to place Σ's as far to the left and Π's as far to the right as possible". This seems to be his closest published approximation to Skolem normal form.

3 Church (1956a), 300n.502; Bochenski (1961), 359. That Aristotle and Aquinas had it straight is pointed out in Quine (1960), 116n.5. For a later statement (c.1903) by Peirce, see 4.464.

*G-1893-5. *Grand Logic*. Selections from this completed but unpublished book are scattered throughout *CP*. (1) Chapter 14, "Second Intentional Logic" (4.80-84), is a brief but interesting note on the logic of quantification (Kneale and Kneale [1962], 433). (2) Peirce defends the doctrine of association according to which the suggestion of some concept B by another concept A is split into two operations, "one leading from A to AB and the other from AB to B" (7.393). This passage is examined in section 7.11 above. (3) Peirce describes a concept as "the living influence upon us of a *diagram* or *icon*" (7.467). (4) An unpublished portion of this book (Ms 410) is the source of the Kempe-like diagram, given in 2.3 above, of the proposition 'Every mother loves some child of hers'.

G-c.1895-1. "That Categorical and Hypothetical Propositions are one in essence, with some connected matters". Selections from this are scattered in *CP*. "In every assertion we may distinguish a speaker and a listener" (2.334). This is related to Peirce's notion that all thinking is dialogical, and to his distinction between the graphist and the interpreter in many of his accounts of EG.

G-1896-3. "A Graphical Method of Logic", a paper presented November 17-18 to the National Academy of Sciences.

*G-1896-6a. "The Regenerated Logic", 3.425-455. This item and the next are parts of a *Monist* review of Schröder [1895]. In this first part there is a valuable passage regarding the benefits to be expected from logical calculi (3.426-429; Bochenski [1961], 279-280); there is here (3.443), as there was in G-1885-3, a statement attempting to justify the use of material implication in formal logic (Bochenski [1961], 313-314); in addition there is the first clear presentation of 'formal implication' (3.445; Bochenski [1961], 354).

*G-1896-6b. "The Logic of Relatives", 3.456-552. In this part of the review of Schröder there is another anticipation of prenex normal form (Church [1956a], 292n.469; Kneale and Kneale [1962], 431-432), and an interesting and useful interpretation of the doctrine of relatives in terms of a subject-predicate classification (Kneale and Kneale [1962], 432). In this article Peirce introduces his entitative graphs.

1896. Peirce invents EG. See Ms 498, Ms Am 806*, Ms 500, and Ms L 477.

From this time on there are a great many manuscripts concerned with logical graphs. Only a few of the more interesting items are included in this list.

G-1898-1. Several sets of lectures are included under this number. One set, delivered at the Cambridge Conferences of 1898, had the title "Reasoning and the Logic of Things"; another was apparently delivered under the title "Detached Ideas on Vitally Important Topics". In a first lecture (Ms 436) Peirce mentions the need for a brief and simple introduction to the logic of relatives and exact logic in general; as originally planned, the lecture would have presented that simple introduction while comparing the entitative and existential graphs. (See Peirce to William James, a letter dated December 18, 1897; Ms L 224.)

1900. Peirce to Christine Ladd-Franklin, a letter dated November 9, 1900. Ms L 237. "You ask whether Logical Graphs have any bearing on Non-Relative logic. Not *much, except* in one highly important particular, that they supply an entirely new system of fundamental assumptions to logical algebra". See section 7.22 above.

1900. On December 11, 1900, Peirce writes out the primitive basis for "A New Logical Algebra" which includes, as a primitive rule of inference, the deduction theorem (*LN*, p. 180r). This is an anticipation of the natural deduction systems developed by Gentzen and Jaśkowski. See Chapter 7.22 above.

G-1901-6. Contributions to Baldwin (1902). Many of Peirce's contributions to this dictionary are relevant to his work on logical graphs, and two items are mentioned here. (1) In an article entitled "Symbolic Logic" Peirce writes:

If symbolic logic be defined as logic—for the present only deductive logic—treated by means of a special system of symbols, either devised for the purpose or extended to logical from other uses, it will be convenient not to confine the symbols used to algebraic symbols, but to include some graphical symbols as well [4.372].

(2) In an article entitled "Logic (Exact)", Peirce acknowledges that it is logical algebra which has chiefly been pursued, but he suggests that logical graphs "lead more directly to the ultimate analysis of logical problems than any algebra yet devised" (3.619). This theme is often to be repeated in Peirce's writings.

N-1902-13. In a review of Friedrich Paulsen's *Immanuel Kant: His Life and Doctrine*, Peirce writes: "Accordingly Kant's great engine and distinction is accurate analysis. But absolute completeness of logical analysis is no less unattainable [than] is omniscience. Carry it as far as you please, and something will always remain unanalyzed" (Ms 1454).

*G-c.1902-2. *Minute Logic*. Parts of this uncompleted book are found throughout *CP*. We are concerned with Chapter 3, "The Simplest Mathematics", 4.227-323. In this chapter Peirce again discusses truth-tables (not by this name), giving them a theoretical basis, illustrating with an example (4.260-262; Berry [1952], 158), and using them to define statement connectives (Bochenski [1961], 330-331). He again mentions the single connective for the propositional calculus (4.264) which he first mentioned in G-c.1880-1 (Bochenski [1961], 344-345; Church [1956a], 133n.207). Peirce also reasserts his view that metaphysics should be based on logic (2.121); he emphasizes the part played by observation in mathematical or necessary reasoning (4.233); and he distinguishes between the mathematician and the logician by saying that the mathematician is interested in drawing inferences, the logician in the analysis of inferences (4.239).

G-1903-2. The Lowell Lectures of 1903. In many of these lectures and in a syllabus prepared to supplement them Peirce develops EG in terms of the three parts Alpha, Beta, and Gamma. This material is a major source for the exposition in Chapters 3, 4, and 5 above.

G-1906-2. "Recent Developments of Existential Graphs and their Consequences for Logic", a paper presented April 16-18 to the National Academy of Sciences. 4.573-584 are from it. Peirce applies EG to the problem of the composition of concepts, and argues for the reality of possibilities. This lecture marks an early stage in the development of the tinctures.

G-1905-1c. "Prolegomena to an Apology for Pragmaticism", 4.530-572. This, the third in the *Monist* 1905-1906 series on pragmaticism, was published in October of 1906. It contains Peirce's tinctured existential graphs, discussed in detail in Chapter 6 above.

1909. Peirce to William James, a letter dated January 8, 1909. Ms L 224. Peirce speaks of a proposed *Monist* article in which he will use EG to "show in what logical analysis, i.e. definition, really consists".

1909. From February 16 to February 23 Peirce investigated systems of triadic logic, recording it in his *Logic Notebook*, pp. 340v-344r. He anticipates certain results usually credited to Łukasiewicz and Post (Fisch and Turquette [1966], and Turquette [1967] and [1969]).

1909. In a manuscript entitled "Studies in Meaning", dated from March 26 of 1909, Peirce writes: "My 'Existential Graphs' have a remarkable likeness to my thoughts about any topic of philosophy" (Ms 620).

1911. Ms 670, "Assurance through Reasoning", written in June, introduces the tinctures into an exposition of EG.

1911. Peirce to A. D. Risteen, an uncompleted letter begun December 6, 1911. Ms L 376. Peirce gives a sketch of the history of EG, reaffirms his opinion that all reasoning is dialogical, and points out that the purpose of EG was not to serve as a calculus, but "to facilitate the anatomy, and thereby the physiology of deductive reasonings". This manuscript contains the only reference I have found to a proposed Delta part of EG which would deal with modal logic.

1913. Peirce to F. A. Woods, a letter begun October 14, 1913. Ms L 477. This letter also begins with a short history of EG. Peirce then claims to have found a fallacy in Prolegomena (G-1905-1c) which leads him to give an alternative graph for Fig. 210 of 4.569. This passage is discussed in detail in section 6.22 above.

APPENDIX 2

TABLE OF LOGIC NOTATIONS

English	Entitative Graphs	Existential Graphs	A Peirce Algebra	*Principia* Notation	Polish Notation
P	P	P	P	P	P
Not-P			$\bar{P}$	~P	Np
P and Q		P Q	PQ	P . Q	Kpq
P or Q	PQ		P ψ Q	P v Q	Apq
If P then Q	 Q		P −< Q	P ⊃ Q	Cpq
All F is G			$\Pi_x(\bar{F}_x \psi G_x)$	(∀x) [Fx ⊃ Gx]	ΠxCfxgx
No F is G			$\Pi_x(\bar{F}_x \psi \bar{G}_x)$	(∀x) [Fx ⊃ ~Gx]	ΠxCfxNgx
Some F is G			$\Sigma_x F_x G_x$	(∃x) [Fx . Gx]	ΣxKfxgx
Some F is not G			$\Sigma_x F_x \bar{G}_x$	(∃x) [Fx . ~Gx]	ΣxKfxNgx

APPENDIX 3

EG CONVENTIONS AND RULES

Reference to similar statements in *CP* is given for most of the conventions and rules; references to the unpublished manuscripts are given in the text. Certain Gamma conventions (see e.g. 5.11 and 5.13) and certain changes occasioned by the tinctures (see e.g. 6.2 and 6.3) are omitted from this list.

Conventions

C1. The sheet of assertion in all of its parts is a graph. 4.396, 397.

C2. Whatever is scribed on the sheet of assertion is asserted to be true of the universe represented by that sheet. 4.397.

C3. Graphs scribed on different parts of the sheet of assertion are all asserted to be true. 4.433.

C4. The scroll is the sign of a conditional proposition *de inesse*. 4.401, 435, 437.

C5. The empty cut is the pseudograph; and the cut precisely denies its contents. 4.467.

C6. The scribing of a heavy dot or unattached line on the sheet of assertion denotes the existence of a single, individual (but otherwise undesignated) object in the universe of discourse. And it is always permitted to scribe such a dot or line on the sheet. 4.404, 405, 417, 559, 567.

C7. A heavy line, called a line of identity, shall be a graph asserting the numerical identity of the individuals denoted by its two extremities. 4.406, 444.

C8. A branching line of identity with *n* number of branches will be used to express the identity of the *n* individuals denoted by its *n* extremities. 4.446, 561.

C9. Points on a cut shall be considered to lie outside the area of that cut. 4.407, 450.

C10. The broken cut expresses that the entire graph on its area is logically contingent (non-necessary). 4.410, 515.

C11. For the interpretation of a line of identity which extends from metal to color or from metal to fur, metal takes precedence; that is, the line does not denote the abstraction (represented by the color or fur), but denotes an existing individual to whom the abstraction pertains.

Rules of Transformation

R1. The rule of erasure. Any evenly enclosed graph and any evenly enclosed portion of a line of identity may be erased. 4.492(1), 505.

R2. The rule of insertion. Any graph may be scribed on any oddly enclosed area, and two lines of identity (or portions of lines) oddly enclosed on the same area, may be joined. 4.492(1), 505.

R3. The rule of iteration. If a graph P occurs on SA or in a nest of cuts, it may be scribed on any area not part of P, which is contained by {P}. Consequently, (a) a branch with a loose end may be added to any line of identity, provided that no crossing of cuts results from this addition; (b) any loose end of a ligature may be extended inwards through cuts; (c) any ligature thus extended may be joined to the corresponding ligature of an iterated instance of a graph; and (d) a cycle may be formed by joining, by inward extensions, the two loose ends that are the innermost parts of a ligature. 4.492(2), 506.

R4. The rule of deiteration. Any graph whose occurrence could be the result of iteration may be erased. Consequently, (a) a branch with a loose end may be retracted into any line of identity, provided that no crossing of cuts occurs in the retraction; (b) any loose end of a ligature may be retracted outwards through cuts; and (c) any cyclical part of a ligature may be cut at its inmost part. 4.492(2), 506.

R5. The rule of the double cut. The double cut may be inserted around or removed (where it occurs) from any graph on any area. And these transformations will not be prevented by the presence of ligatures passing from outside the outer cut to inside the inner cut. 4.492(4), 508, 567.

R6. The rule of cut conversion. (a) An evenly enclosed standard cut may be transformed (by being half erased) into a broken cut; and (b) an oddly enclosed broken cut may be transformed (by being filled up) into a standard cut. 4.516.

R7. The rule of modal selectives. From  we can infer 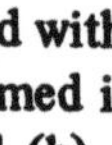 . 4.518.

R8. Any graph, evenly enclosed on metal, may be tinctured with color.

R9. Any graph, oddly enclosed on color, may be tinctured with metal.

R10. (a) Any province tinctured in color can be transformed into a province of metal having a border in that color; and (b) a province tinctured in metal but with a border in some tincture of color can be transformed into a province of that color.

APPENDIX 4

COMPLETENESS AND CONSISTENCY

1 *Alpha*

A. *Completeness.* To show that Alpha is deductively complete, it is sufficient to prove, as theorems and derived rules of Alpha, the axioms and rules of inference of some other formulation of the propositional calculus which is known to be complete. As an example of such a formulation consider system P of Church (1956a), 149, whose primitive symbols include only propositional variables, brackets, and the two operator symbols ⊃ and ~. There are three formation rules ('wff' and 'wf' abbreviate 'well-formed formula' and 'well-formed', respectively):

- i. A variable standing alone is a wff.
- ii. If P is a wff, then ~P is a wff.
- iii. If P and Q are wffs, then [P ⊃ Q] is a wff.

The axioms of system P are given by the three following schemata:

P1. [P ⊃ [Q ⊃ P]]
P2. [[P ⊃ [Q ⊃ R]] ⊃ [[P ⊃ Q] ⊃ [P ⊃ R]]]
P3. [[~P ⊃ ~ Q] ⊃ [Q ⊃ P]]

The single rule of inference is *modus ponens*:[1]

R1. From [P ⊃ Q] and P to infer Q.

Any wff of system P can be translated into graphical notation by the two conventions that (1) every wf part of the form [P ⊃ Q] is to be replaced by a graph of the form , and (2) every wf part of the form ~P is to be replaced by a graph of the form . Applying these conventions to the axioms and rule of system P gives us the following graphs (the suffix '(a)' marks an Alpha translation of a wff of system P):

1 Since P employs axiom schemata rather than axioms, no rule of substitution is necessary. This is true also for the system ML to be introduced in section 2, B.

P1(a). 

P2(a). 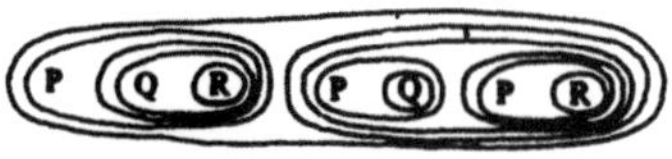

P3(a). 

R1(a). From  and P to infer Q.

Now a proof in system P is a finite sequence of wffs, each of which is either an axiom or the result of a use of *modus ponens*. Hence, just as the axioms and rule of system P have been translated into EG, so each proof in P can be rewritten step by step into graphical notation. And the result of this translation will be a proof in EG *if* we can obtain proofs of P1(a), P2(a), P3(a), and R1(a) in Alpha. Alpha proofs of R1(a) and P2(a) have already been given in Chapter 3, section 3, and proofs of P1(a) and P3(a) now follow.

P1(a). | 

1.  R5.

2. 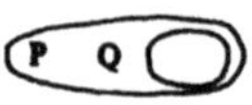 1, by R2.

3.  2, by R3.

Now add double cuts by R5.

P3(a). 

1.  R5.

2. 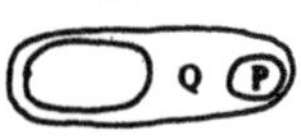 1, by R2.

3.  2, by R3.

4.  3, by R5 twice.

This completes the proof that Alpha is deductively complete.

B. *Consistency*. By 'elementary graph' is meant a graph according to C1 or C2, that is, a graph which contains no cuts and is not a juxtaposition of other graphs. By 'tautology' is meant a graph whose value, according to the valuation procedure sketched in 3.2 above, is 1 (true) for all possible assignments of values to its elementary graphs. In order to prove that Alpha is consistent, we prove first that every theorem of Alpha is a tautology. To do this it is sufficient to show that the single axiom of Alpha is a tautology, and that the rules of transformation preserve tautologies in the sense that, when the rules are applied to a tautology as premiss, the conclusion must also be a tautology.[2]

Before even this is done, however, it is convenient to prove the five following lemmas. Keep in mind that an area enclosed by n number of cuts is called the *nth* area of a given nest of cuts (see 3.1 and Glossary). Further, by 'value of the cut K' is meant the value of the enclosure whose outermost cut is K.

Lemma 1. Let α and β be areas of the nest of cuts N, such that $\alpha \supseteq \beta$. If the value of β has no effect on the value of α, then the value of β has no effect on the value of N.

Proof: This follows from the fact that, according to the valuation procedure, the calculation of the value of a nest of cuts begins from inside the nest and proceeds outwards. And in this calculation the value of each area is used exactly once.

Lemma 2. To erase a graph from an area α can change the value of α from 2 to 1 but not from 1 to 2.

Proof: That the erasure of a graph P from α cannot change the value of α from 1 to 2 follows from the way in which the valuation procedure assigns a value to juxtaposition (conjunction), and from the stipulation that the blank has the value 1 (which is relevant in case the erasure empties α). But the erasure can change the value of α from 2 to 1, and it will do so in case P = 2 and no other graph on α has the value 2.

Lemma 3. To insert a graph onto an area α can change the value of α from 1 to 2 but not from 2 to 1.

Proof: That the insertion of P onto α cannot change the value of α from 2 to 1 follows from the way juxtapostion is evaluated. But such an insertion can change the value of α from 1 to 2, and will do so in case P = 2.

Lemma 4. To erase a graph from an area β which is contained by an area α and enclosed by two more cuts than α can change the value of α from 2 to 1 but not from 1 to 2.

Proof: To fix our ideas, consider Fig. 1 below, and suppose that α is enclosed by n cuts, β by n + 2 cuts.

We distinguish two cases.

Case 1. The value of α is 1. By lemma 2, to erase a graph from β will

[2] A method of proof different from the one used here is sketched in Roberts (1964), 118f. There the idea is to obtain suitable translations of the Alpha rules as derived rules in a system of logic already known to be consistent. The more direct approach of this appendix seemed interesting in its own right.

Fig. 1

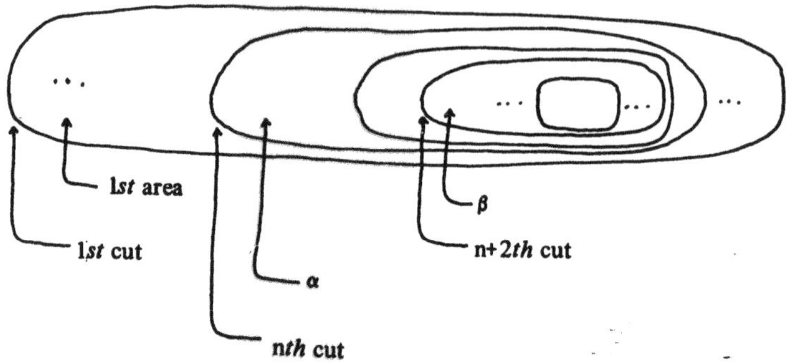

either make no change in the value of β or will change it from 2 to 1. If the former, then, of course, no change will occur in the value of α. And if the latter, again no change will be effected in the value of α; for when the value of β is 1—no other value changes being made in the nest of cuts—, straightforward calculation shows that α must retain the value 1. Hence erasures performed on β cannot change the value of α from 1 to 2.

Case 2. The value of α is 2. Then any change in the value of α would be a change from 2 to 1, and this change can be effected by a value change on β. For suppose the value 2 of α is a result of the value 2 of the n + 1*th* cut; then the value of the n + 1*th* area must be 1. To change the value of β from 2 to 1 will change all of this, as straightforward calculation will show.

Lemma 5. To insert a graph onto an area β which is contained by α and enclosed by two more cuts than α can change the value of α from 1 to 2 but not from 2 to 1.

Proof: Consider Fig. 1 again, with α and β identified as in lemma 4. We distinguish two cases.

Case 1. The value of α is 2. By lemma 3, to insert a graph onto β will either make no change in the value of β, or will change it from 1 to 2. If the former, then no change will be effected in the value of α. If the latter, then again no change will occur in the value of α; for if the value of β changes from 1 to 2—no other value changes being made in the nest of cuts—, straightforward calculation shows that α must retain the value 2. Hence insertions onto β cannot change the value of α from 2 to 1.

Case 2. The value of α is 1. Then any change in the value of α would be a change from 1 to 2, which can be effected by an insertion onto β, as the reader can verify for himself.

We now address ourselves to the axiom and rules of Alpha.

The Alpha axiom is a tautology. That this is so, that the blank SA or any of its parts is a tautology, is not a matter of calculation, but of stipulation. This accords with the practice of Peirce, who took SA to represent true propositions,[3] and it is reasonable enough if SA is taken simply as a blank,

3 Recall the discussion leading up to C1 in Chapter 3. Cf. 4.396, 431, 512, and 567.

since to assert nothing is not to assert something false.

R1 preserves tautologies.

Proof: We proceed by mathematical induction on n, where 2n is the number of cuts within which the erasure takes place (n = 0, 1, 2, . . .).

Case 1. Erasure is applied to a graph which is evenly enclosed within zero cuts (n = 0), that is, erasure is applied to a graph which is scribed unenclosed on SA. By lemma 2, the value of the total graph cannot be changed from 1 to 2 by this operation; hence, if the premiss is a tautology, so is the conclusion.

Case 2. Given some nest of cuts, the hypothesis of induction is that erasure preserves tautologies when applied within 2k of those cuts (n = k). We have to show that under this assumption, erasure also preserves tautologies when applied within 2(k+1) cuts (n=k+1). Now according to lemma 2, to say that erasure preserves tautologies when applied within 2k cuts is to say that to change the value of the 2*kth* area from 2 to 1 will not change a tautology into a non-tautology. It follows that any operation which can produce this and no other change on the 2*kth* area must preserve tautologies.[4] But according to lemma 4, the application of erasure to the 2(k + 1)*th* area is just such an operation. Hence, erasure preserves tautologies when applied within 2(k + 1) cuts.

This completes the induction, and we may therefore conclude that erasure preserves tautologies when applied within any even number of cuts. That is, R1 preserves tautologies.

R2 preserves tautologies.

Proof: We proceed by mathematical induction on n, where 2 n + 1 is the number of cuts within which the insertion takes place (n = 0, 1, 2, . . .).

Case 1. Insertion is applied to the first area of a nest of cuts (n = 0). Now if the nest of cuts has the value 1, the value of its first area must be 2; and by lemma 3, insertion cannot change this value. Hence if the premiss is a tautology, so is the conclusion.

Case 2. Given some nest of cuts, the hypothesis of induction is that insertion preserves tautologies when applied within 2k + 1 of those cuts (n = k). We have to show that under this assumption, insertion also preserves tautologies when applied within 2(k + 1) + 1 cuts (n = k + 1). Now according to lemma 3, to say that insertion preserves tautologies when applied within 2k + 1 cuts is to say that to change the value of the 2k + 1*th* area from 1 to 2 will not transform a tautology into a non-tautology. It follows that any operation which can produce only this change on the 2k + 1*th* area must preserve tautologies. But according to lemma 5, the application of insertion within two additional cuts—that is, within 2(k + 1) + 1 cuts—is just

4 We exclude simultaneous erasures from the 2*kth* area and from areas evenly enclosed within fewer than 2k cuts. The operations of EG apply to one area at a time.

such an operation. Hence, insertion preserves tautologies when applied within $2(k + 1) + 1$ cuts.

This completes the induction, and we may therefore conclude that insertion preserves tautologies when applied within any odd number of cuts. That is, R2 preserves tautologies.

It is convenient to treat R3 and R4 together.

R3 and R4 preserve tautologies.

Proof: Let N be a nest of cuts in which the operations are imagined to take place. Let α be the area of the original or remaining occurrence of the graph which is to be iterated or deiterated, and let β be the area onto which the graph is iterated by R3 or from which it is deiterated by R4. It follows that $\alpha \supseteq \beta$. Two cases are distinguished.

Case 1. The graph P to be iterated or deiterated has the value 1. Regardless of the value of β before the use of R3 or R4, it is clear from the way juxtaposition is evaluated that the insertion of P onto β or its removal from β cannot change the value of β, and therefore cannot change the value of N. Hence, if the premiss is a tautology, so is the conclusion.

Case 2. The graph P to be iterated or deiterated has the value 2. Then the value of α is 2 regardless of the value of β before and after the use of R3 or R4, and it follows from this by lemma 1 that the value of β has no effect on the value of N. Hence if the premiss is a tautology, so is the conclusion.

From cases 1 and 2 it follows that R3 and R4 both preserve tautologies.

R5 preserves tautologies. This follows immediately from the valuation procedure as applied to the cut. Namely, for any graph P, P and ((P)) always have the same value.

We summarize our results so far in a sixth lemma:

Lemma 6. Every theorem of Alpha is a tautology.

Among the several senses of consistency that may be distinguished we define two: (a) A system of logic S is said to be *consistent with respect to* a given transformation by which each expression P is transformed into an expression P′, if there is no expression P such that both P and P′ are theorems of S. (b) A system S is said to be *absolutely consistent* if not all of its expressions are theorems (see Church [1956a], 108).

Corollary 1. Alpha is consistent with respect to the transformation of P into (P).

Proof: By the definition of a tautology and the way in which the cut is evaluated, not both P and (P) can be tautologies. Hence, by lemma 6, not both P and (P) can be theorems of Alpha.

Corollary 2. Alpha is absolutely consistent.

Proof: The empty cut is not a tautology (in fact, it always has the value 2), and therefore by lemma 6 is not a theorem of Alpha.

2 *Beta*

A. *Consistency.* Any Beta graph can be transformed into an associated Alpha graph (aag) by a two-step procedure: first, delete all the lines of identity in the Beta graph; then, in the resulting graph, replace each spot by an elementary Alpha graph according to this rule: different occurrences of the same spot are to be replaced by the same elementary Alpha graph, and different spots are to be replaced by different elementary Alpha graphs.

Lemma 1. Every Beta theorem has a tautology as aag.

Proof: The single axiom of Beta has a tautology as aag. This is immediate, since the Beta axiom is the unenclosed line of identity, whose aag is the blank SA.

Furthermore, the Beta rules of transformation preserve the property of having a tautology as aag, that is, if the premiss or premisses of the rule have this property, then the conclusion does also. In the case of R5, this is immediate, since the only difference between the aag of the premiss and the aag of the conclusion will be the presence in one, but not the other, of a double cut; and this difference is fully accounted for by the Alpha version of R5, which was proved to preserve tautologies in the preceding section. It is also immediate in the case of those applications of the first four rules which deal exclusively with the line of identity, since in this case the aag of the premiss is identical to the aag of the conclusion, so that if one is a tautology, so is the other.

All other applications of the first four rules will involve cuts or spots or both as well as lines. Now reduction to an aag eliminates all lines of identity, but apart from this it changes nothing with respect to the structure of a nest of cuts: the only remaining difference is that where the Beta graph has a spot on a given area of the nest, the aag has an elementary graph. Furthermore, the operations upon spots and cuts that are permitted by the Beta rules are precisely the same operations that are permitted by the Alpha rules. This is obvious, since apart from the clauses in the Beta rules which pertain solely to the lines of identity, the Beta rules are identical to the Alpha rules. It follows therefore that the difference between the aag of the premiss and the aag of the conclusion is fully accounted for by the Alpha version of these Beta rules; and since the Alpha rules preserve tautologies (lemma 6 of the preceding section), if the aag of the premiss is a tautology, then the aag of the conclusion is also.

This completes the proof of lemma 1.

Now by the definition of a tautology and the way in which the cut is evaluated, it is clear that if a Beta graph P has a tautology as aag, its negation Ⓟ does not. Hence by lemma 1 not both P and Ⓟ can be theorems. We therefore conclude the following:

Corollary 1. Beta is consistent with respect to the transformation of P into Ⓟ.

Corollary 2. Beta is absolutely consistent.

B. *Completeness.* To show that Beta is deductively complete it is sufficient to prove, as theorems and derived rules of Beta, the axioms and rules of inference of some other formulation of the first order functional calculus which is known to be complete. As an example of such a formulation consider the system ML of Quine (1955), 88. It is convenient for our purposes to treat ML as an extension of system P which was used in section 1 to prove Alpha completeness.[5] Namely, we add to the primitive symbols of system P individual variables, functional variables, and parentheses. There are two additional formation rules:

iv. If F is an n-ary functional variable, and $x_1, x_2, \ldots, x_n$ are individual variables, then $F(x_1 x_2 \ldots x_n)$ is a wff.

v. If P is wf and x is an individual variable, then (x)P is wf.

An individual variable x is *bound* in a wff P if it occurs in a wf part of P of the form (x)Q; otherwise it is *free* in P. If P has n free individual variables, $x_1, x_2, \ldots, x_n$, the result of prefixing to P the n quantifiers $(x_1), (x_2), \ldots, (x_n)$ will be called the *closure* of P. We say 'the' closure of P, in spite of the fact that the quantifiers may be prefixed to P in n! distinct ways; for all these n! wffs are equivalent. The sign '$\vdash$ P' means that the closure of P is a theorem. There are five axioms or principles of ML, but since one of them is *modus ponens*, which is a rule of system P and for which an EG proof has already been given, we need consider only the following four:

*100. If P is tautologous, $\vdash$P.

*101. $\vdash (x)[P \supset Q] \supset [(x)P \supset (x)Q]$.

*102. If x is not free in P, $\vdash P \supset (x)P$.

*103. If P′ is like P except for containing free occurrences of y wherever P contains free occurrences of x, then $\vdash (x)P \supset P'$.

To translate these axioms into graphical notation, we supplement the two conventions of section 1, A above: (3) every wf part of the form (x)P, where x is free in P, is to be replaced by a graph of the form  ; (4) every wf part of the form (x)P, where x is not free in P, is to be replaced by

5 This is not the way Quine treated it. He did not develop the propositional calculus in an axiomatic fashion, and he did not use ⊃ and ~ as primitive symbols. Nevertheless, the use of Quine's system is particularly appropriate, because Quine has specified his axioms in such a way as to ensure that the theorems contain no free individual

a graph of the form , where P has no unoccupied hooks. The hole drawn at the innermost end of the line of identity indicates that that end of the line is not attached to the hook of any spot.[6] We use this sign only when the proximity of a line and a spot might give rise to confusion.

Let P be a Beta graph containing any number of hooks, each hook filled with a line of identity, the lines variously connected and enclosed. By the *n-th degree closure* of P is meant the graph of Fig. 1, formed by enclosing P in two cuts, if necessary, and attaching a once enclosed line to n distinct

Fig. 1

Fig. 2

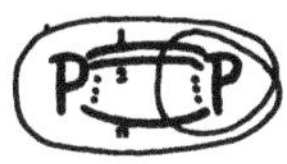

Fig. 3

lines of P. If the graph to be closed is an enclosure, the additional two cuts are superfluous. For example, where P is defined as above, Fig. 3 is the n-th degree closure of Fig. 2. (The hooks are placed on facing sides of the P's to avoid crossing lines of identity.)

There may be more than one n-th degree closure of the same graph, but these differences do not affect our present discussion. When our remarks apply to all degrees of closure, we shall speak of 'closure' or 'the closure' without specifying any degree.

The closures of many Beta graphs can be obtained as Beta theorems by variations of the following proof schema, to be referred to by the abbreviation 'CPS' for 'closure proof schema'. We prove that Fig. 3 above is a Beta theorem. Note that what is inserted in step 2 of the proof is not necessarily the graph P, but the graph of Fig. 4, which we shall call the *n-th degree extension* of P, abbreviated P_n; it is formed from P by extending outwards,

Fig. 4

beyond all cuts of P, the n lines with respect to which closure is to be obtained. It is not excluded that some of these n lines are already thus extended in P; if all of them are, P is the same graph as P_n.

1. R5.

---

variables. It was pointed out above in Chapters 4 and 7 that formulas with free individual variables cannot be expressed in EG. In using ML to prove Beta completeness I follow the lead of Zeman (1964), but my account differs from Zeman's in the graphical translations and proofs of several ML axioms.

6 Peirce did not use a special sign for vacuous quantification, but he knew it could occur in EG, as his Beta axiom and rules R3 and R4 make clear. In Ms 478, p. 168, he makes it explicit in the following theorem. "A branch of a Ligature extending into the Area of a Cut without then being connected with any Hook is of no effect".

2.  1, by R2.

3.  2, by R3.

4.  3, by R3(b).

Now use R3(c) to join the lines. Upcoming proofs in the style of CPS will be abbreviated, usually combining in one step the iteration by R3, the inward extension of lines by R3(b), and the joining of lines by R3(c).

Consider first *100: If P is tautologous, ⊢ P. A 'tautologous' wff of ML is a wff having the form of a tautology, that is, a wff which can be obtained by replacing the propositional variables of a tautology of system P by wffs of ML according to the rule that different occurrences of the same variable are to be replaced by the same wff, and different variables are to be replaced by different wffs. Similarly, any Alpha graph P (whether a tautology or not) can be transformed into an associated Beta graph (abg), symbolized P*, by replacing each of the elementary graphs of P with Beta graphs (which may be spots or may contain spots, whose hooks must each be filled with a heavy dot) according to this rule: different occurrences of the same elementary graph are to be replaced by the same Beta graph, and different elementary graphs are to be replaced by different Beta graphs.

Our proof of *100 in Beta makes use of the fact that any wff P of system P can be translated into an Alpha graph P(a) by the conventions of section 1 above; and P(a) can be transformed into an abg P(a)* by the procedure just described. Now it was proved in section 1 that if all the steps in a system P proof of a theorem P are translated into graphs, the result is an Alpha proof of the graph P(a). Each step of such a proof is an EG translation of an axiom of system P—either P1(a), P2(a), or P3(a)—or an EG translation of the result of *modus ponens*. Transform each graph of such an Alpha proof of P(a) into the closure of its abg, and the result will be a Beta proof of the closure of P(a)*, *if* we can prove as Beta theorems the closures of P1(a)*, P2(a)*, and P3(a)*, and *if* we can prove as a derived rule of Beta that *modus ponens* preserves closure, in the sense that if the rule is applied to premisses which are closed, then the conclusion is closed. It would follow from this that the closure of any abg of every Alpha theorem is a Beta theorem,[7] and since the EG translation of any tautology of system P is an Alpha theorem, this would complete the proof of the following graphical version of *100:

100. If P is a tautology (of system P), the closure of P(a) is a Beta theorem.

We employ the method of the CPS, selecting a degree of closure appro-

[7] The method is, in effect, that of mathematical induction with respect to the number of steps in the proof of the closure of any Alpha theorem.

priate to the structure of each axiom. Proofs for closures of any degree would be strictly similar.

1. The second degree closure of P1(a)*, given in Fig. 5, is a Beta theorem.

Fig. 5 

1.  R5, R2.

2.  1, by R3, as in CPS.

Now add double cuts by R5.

2. The third degree closure of P2(a)*, given in Fig. 6, is a Beta theorem.

Fig. 6 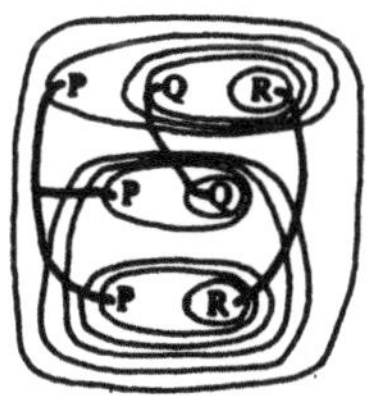

1. 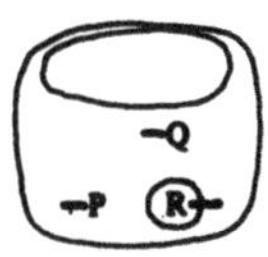 R5, R2.

2. 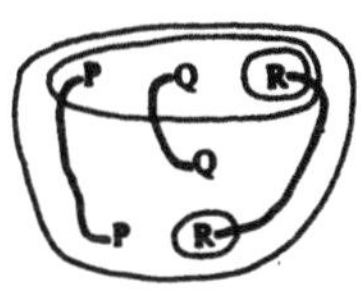 1, by R3, as in CPS.

3. 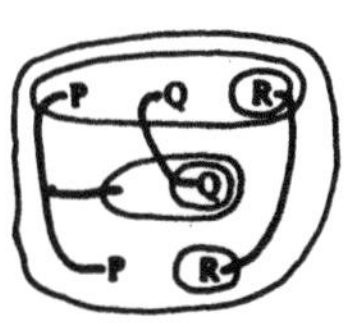 2, by R5 and then R3(a) and (b).

Now use R3 to attach–P to the loose end of the line, and add three sets of double cuts by R5.

3. The second degree closure of P3(a)*, given in Fig. 7, is a Beta theorem.

Fig. 7 

1.  R5, R2.

2.  1, by R3, as in CPS.

Now add two sets of double cuts by R5.

4. *Modus ponens* preserves closure. That is, from the graph of Fig. 8 we can infer the graph of Fig. 9.

Fig. 8

Fig. 9

1.  Premiss.

2.  Premiss.

3.  1, by R5.

4.  3, by R1.

5.  4, by R4(b).

6.  5, 2 by R4.

This completes the proof of *100.

*101 may be translated into the graph of Fig.10, under the assumption that only the variable x occurs free in P and Q. For cases in which other

Fig. 10

variables occur free, the method of CPS (applied in the step 1 use of R3) suffices to prove closure.

1.  R5, R2, R3.

2.  1, by R5.

3.  2, by R1 (breaking the line on even, at the arrow drawn in step 2) and R4(b).

Now add double cuts by R5.

*102 may be translated into the graph of Fig. 11, under the assumption

Fig. 11

that no variable occurs free in P. Otherwise, the method of CPS (in the step 1 use of R3) will give closure.

1.  R5, R2, R3.

2.  1, by R3(b).

Now add double cuts by R5.

*103 may be translated into the graph of Fig. 12, under the assumption that only x occurs free in P. Otherwise, the method of CPS (in the step 2 use of R3) will give closure.[8]

Fig. 12

1. 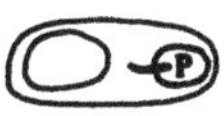 R5, R2.

2.  R3.

If it is desired to force the reading of the once enclosed line as a prefix to the scroll itself, simply lengthen the line by R3(a) and add double cuts by R5, obtaining Fig. 13.

Fig. 13

This completes the proof of Beta completeness. That Beta also contains the axioms of identity was proved in Chapter 4, section 3.

8 The prime notation in the ML version of *103 signifies that the individuals denoted by the individual variables may be different; this is accomplished in the graphical version by the use of unconnected lines of identity.

APPENDIX 5

GLOSSARY

Area of a cut. The space within or enclosed by a cut.

Blot. One form of the pseudograph; a cut whose area is completely blackened: 

Broken cut. "A cut with many little interruptions aggregating about half its length" (4.410).

Close of a cut. Same as 'area'.

Cut. "A self-returning finely drawn line" (4.414). A finely drawn closed line. Used as the sign of negation (it precisely denies its contents) and, when empty, as the pseudograph. Sometimes spoken of as an actual incision in SA (4.556). See 'verso'.

Deiteration. The erasure of a graph of which at least two instances occur, according to R4.

Double cut. A scroll with nothing in the outer area except, perhaps, lines of identity extending from outside the outer cut to inside the inner cut.

EG. Abbreviation for 'the system of existential graphs'.

Enclosure. A cut (4.399, 4.414) or a scroll (4.437) taken together with its contents.

Endoporeutic. Proceeding from the outside inward.

Entire graph. Everything that is scribed on SA.

Entitative. Defined by Peirce as "pertaining to actually existing objects .... Entitative being, existence, real being; opposed to intentional or objective being" (Ms 1167, p. 17). See also the definition of this term in the *Century Dictionary.*

Erasure. The removal of a graph instance from some area.

Evenly enclosed. Enclosed within 2n number of cuts, where n = 0 or any of the positive integers.

Existential. Defined in the *Century Dictionary* as "Of, pertaining to, or consisting in existence; ontological".

First area. That area in a nest of cuts which is enclosed by precisely one cut.

First close. Same as 'first area'.

Graph. A proposition, an expression of any possible state of the universe. In Peirce's terminology, a graph is a type or universal, and is to be distinguished from a graph-instance.
Graph-instance. An occurrence of a graph. In Peirce's terminology, a graph-instance is a token, and is to be distinguished from a graph.
Graph-replica. Same as 'graph-instance'.
Heavy dash. A sign of the individual: ▬
Heavy dot. Another sign of the individual: ●
Hook. The place (usually not indicated in any special way) on the periphery of a spot to which a line of identity can be attached.
Inner area. The area within the inner cut of a double cut or scroll.
Inner close. Same as 'inner area'.
Inner cut. That cut of a double cut or scroll which is itself enclosed by another cut.
Insertion. The scribing of a graph upon some area.
Instance. See 'graph-instance'.
Iteration. The insertion of a graph of which at least one instance already occurs on some area, according to R3.
Ligature. The totality of all the lines of identity that join one another.
Line of identity. A heavily drawn line ▬ which served as the quantifier and as the sign of the individual for Peirce.
n-th area. That area of a nest of cuts which is enclosed by precisely n number of cuts.
Nest of cuts. A series of cuts each enclosing the next one.
Node. A thickened part of a line of identity. Peirce, for a time, drew the line of identity with a node at each extremity, as •–•.
Oddly enclosed. Enclosed within 2n + 1 number of cuts, where n = 0 or any of the positive integers.
Outer close. Same as 'first area'.
Outer cut. That cut of a double cut or scroll (or nest of cuts generally) which is not enclosed, but which encloses the other cut (or cuts) within itself.
Partial graph. Any graph scribed in the presence of other graphs.
Peg. Same as 'hook'. See 4.621.
Phemic sheet. A term introduced in 1906 for the broadened interpretation of the sheet on which graphs are to be scribed (4.553ff). See end of section 6.1 and sections 6.2 and 6.22.
Place of a cut or graph. The area on which a cut is made or a graph is scribed. Symbolized by braces, so that {P} denotes the place of P.
Pseudograph. The graphical expression of an absurdity, an 'always false' proposition. Usually symbolized by the blot or the empty cut.
Recto. The surface (unenclosed) of SA.

Relative term. A rhema of more than one blank.

Replica. Early term for 'instance'. See 4.395n.1.

Rhema. Peirce's term for what is now sometimes called a 'propositional function'. Peirce defines it as "a blank form of proposition produced by such erasures as can be filled, each with a proper name, to make a proposition again" (4.438). See Section 7.121 above.

SA. Abbreviation for 'the sheet of assertion'.

Scribe. To write or draw or otherwise place a graph on SA. "Since it is sometimes awkward to say that a graph is *written* and it is sometimes awkward to say it is *drawn*, I will always say it is *scribed*" (Ms 450, p. 8 verso).

Scroll. Two cuts, one within the other, with or without graphs scribed on the inner or outer areas. Thus, the double cut is a form of the scroll. The scroll is the sign of material implication.

Second area. That area of a nest of cuts which is enclosed by precisely two cuts. The inner area of a double cut.

Second close. See 'inner close'.

Selective. A letter used to designate an individual, used by Peirce as an abbreviation. See 4.460. A sign resembling a heavy dot or dash, or extended into a line of identity, attached to broken cuts or other graphs to indicate "the state of information at the time of learning that graph to be true" (4.518).

Sep. Same as 'cut'.

Sheet of assertion. The surface upon which existential graphs are scribed. Abbreviated 'SA'.

Spot. An unanalyzed expression of a rhema (4.441).

Total graph. The entire graph together with SA.

Universe of discourse. The domain of objects represented by SA. "DeMorgan introduced the term on November 6, 1846. Exact logic dates from that day" (Ms 450, p. 7).

Verso. The reverse side of SA. The side of SA exposed when a cut, understood as an incision, is made in SA and the excised piece turned over. The place of graphs that are denied (4.556); for a brief time in 1906, the representation of a universe of possibilities (4.581).

BIBLIOGRAPHY

Aristotle

1941 *The Basic Works of Aristotle*, edited by Richard McKeon (Random House, New York).

Baldwin, James Mark

1901-1902 (Editor) *Dictionary of Philosophy and Psychology*, 3 vols. (The Macmillan Co., New York), Vol. I, 1901; vol. II, 1902.

Berry, George D. W.

1952 "Peirce's Contributions to the Logic of Statements and Quantifiers", Wiener and Young (1952), 153-165.

Bochenski, I. M.

1961 *A History of Formal Logic*, translated and edited by Ivo Thomas (University of Notre Dame Press, Notre Dame, Indiana).

Century Dictionary

1889-1891 *The Century Dictionary and Cyclopedia*, first edition in 6 vols. (The Century Co., New York). In 1909 *The Century Dictionary Supplement* appeared in 2 vols., and in 1911 the complete dictionary was published with revisions and additions. Peirce was a contributor to this work.

Church, Alonzo

1955 "Intention", Runes (1965), 148.

1956a *Introduction to Mathematical Logic*, Vol. I (Princeton University Press, Princeton, N.J.).

1956b "Propositions and Sentences", *The Problem of Universals* (University of Notre Dame Press, Notre Dame, Indiana).

Copi, Irving

1961 *Introduction to Logic*, second edition (The Macmillan Company, New York).

1954 *Symbolic Logic*, first edition (The Macmillan Company, New York).

Feibleman, James K.

1946 *An Introduction to Peirce's Philosophy, Interpreted as a System*, with a foreword by Bertrand Russell (The Hauser Press, New Orleans).

Fisch, Max H. and Jackson I. Cope

1952 "Peirce at the Johns Hopkins University", Wiener and Young (1952), 277-311.

Fisch, Max and Atwell Turquette

1966 "Peirce's Triadic Logic", *Transactions of the Charles S. Peirce Society* 2, 71-85.

Gallie, W. B.
1952 *Peirce and Pragmatism* (Penguin Books, Middlesex).
Gardner, Martin
1958 *Logic Machines and Diagrams* (McGraw-Hill Book Company, Inc., New York).
Gentzen, Gerhard
1934-1935 "Untersuchungen über das logische Schliessen", *Mathematische Zeitschrift* 39, 176-210, 405-431.
Goudge, Thomas A.
1950 *The Thought of C. S. Peirce* (University of Toronto Press, Toronto).
Hintikka, Jaako
1959 "Existential Presuppositions and Existential Commitments", *The Journal of Philosophy* 56, 125-137.
Hughes, G. E. and M. J. Cresswell
1968 *An Introduction to Modal Logic* (Methuen and Co. Ltd., London).
Jaśkowski, Stanisław
1934 *On the Rules of Suppositions in Formal Logic* (= *Studia Logica*, No. 1) (Warsaw).
Kempe, Alfred Bray
1886 "Memoir on the Theory of Mathematical Form", *Philosophical Transactions of the Royal Society of London* 177, 1-70.
1887 "Note to a Memoir on the Theory of Mathematical Form ('Phil. Trans.' (vol. 177), p. 1)", *Proceedings of the Royal Society* 42, 193-196.
1897 "A Correction and Explanation", *The Monist* 7, 453-458.
Kneale, William and Martha Kneale
1962 *The Development of Logic* (Oxford University Press, London).
Leblanc, Hughes and Theodore Hailperin
1959 "Nondesignating Singular Terms", *Philosophical Review* 68, 239-243.
Leonard, Henry
1956 "The Logic of Existence", *Philosophical Studies* 7, 49-64.
Lewis, C. I.
1960 *A Survey of Symbolic Logic*, a corrected republication of the original work with the omission of chapters 5 and 6 (Dover Publications, Inc., New York).
Łukasiewicz, Jan
1957 *Aristotle's Syllogistic, from the Standpoint of Modern Formal Logic*, second edition, enlarged (Oxford University Press, Oxford).
Moore, Edward C. and Richard S. Robin
1964 (Editors) *Studies in the Philosophy of Charles Sanders Peirce*, second series (The University of Massachusetts Press, Amherst, Mass.).
Murphey, Murray G.
1961 *The Development of Peirce's Philosophy* (Harvard University Press, Cambridge, Mass.).
Peirce, Benjamin
1881 "Linear Associative Algebra", with notes and addenda by C.S. Peirce, *American Journal of Mathematics* 4, 97-229.
Peirce, Charles Sanders
n.d. The Charles S. Peirce Papers. Manuscript collection in the Houghton Library, Harvard University, Cambridge, Mass.
1883 (Editor) *Studies in Logic, By Members of the Johns Hopkins University* (Little, Brown and Company, Boston).

1931-1958 *Collected Papers of Charles Sanders Peirce*, 8 vols. Edited by Charles Hartshorne, Paul Weiss, and Arthur Burks (Harvard University Press, Cambridge, Mass.).

1953 *Charles S. Peirce's Letters to Lady Welby*, edited by Irwin C. Lieb (Whitlock's Inc., New Haven, Conn.).

Post, Emil

1921 "Introduction to a General Theory of Elementary Propositions", *American Journal of Mathematics* 43, 163-185.

Prior, A. N.

1962 *Formal Logic*, second edition (Oxford University Press, London).

Quine, Willard Van Orman

1934 "Review of the Collected Papers of Charles Sanders Peirce, Volume 4: *The Simplest Mathematics*", *Isis* 22, 551-553.

1955 *Mathematical Logic*, revised edition (Harvard University Press, Cambridge, Mass.).

1959 *Methods of Logic*, revised edition (Holt, Rinehart and Winston, New York).

1960 *Word and Object* (The Technology Press of the Massachusetts Institute of Technology and John Wiley & Sons, Inc., New York).

Roberts, Don D.

1963 "The Existential Graphs of Charles S. Peirce", an unpublished doctoral dissertation (The University of Illinois, Urbana).

1964 "The Existential Graphs and Natural Deduction", Moore and Robin (1964), 109-121.

Robin, Richard S.

1967 *Annotated Catalogue of the Papers of Charles S. Peirce* (The University of Massachusetts Press, Amherst, Mass.).

1971 "The Peirce Papers: A Supplementary Catalogue", *Transactions of the Charles S. Peirce Society* 7, 37-57.

Ruby, Lionel.

1960 *Logic, an Introduction*, second edition (J. B. Lippincott Company, Chicago).

Runes, Dagobert D.

1955 *Dictionary of Philosophy* (Littlefield, Adams & Co., Ames, Iowa).

Schafter, Thomas Lee

1968 "Existential Graphs", an unpublished masters thesis (Colorado State University, Fort Collins, Colo.).

Schröder, Ernst

1895 *Vorlesungen über die Algebra der Logik (exacte Logik)*, vol. 3, *Algebra und Logik der Relative*, part 1 (Leipzig).

Sheffer, H. M.

1913 "A Set of Five Independent Postulates for Boolean Algebra, with Application to Logical Constants", *Transactions of the American Mathematical Society* 14, 481-488.

Turquette, Atwell

1967 "Peirce's *Phi* and *Psi* Operators for Triadic Logic", *Transactions of the Charles S. Peirce Society* 3, 66-73.

1969 "Peirce's Complete System of Triadic Logic", *Transactions of the Charles S. Peirce Society* 5, 199-210.

Wells, Rulon
1959 Review of Berry (1952), *The Journal of Symbolic Logic* 24, 209-211.
Whitehead, Alfred North
1948 *An Introduction to Mathematics*, revised edition (Oxford University Press, London).
Whitehead, Alfred North and Bertrand Russell
1927 *Principia Mathematica*, 3 vols., second edition (Cambridge University Press, Cambridge).
Wiener, Philip P. and Frederic H. Young
1952 (Editors) *Studies in the Philosophy of Charles Sanders Peirce* (Harvard University Press, Cambridge, Mass.).
Zeman, J. Jay
1964 "The Graphical Logic of C. S. Peirce", an unpublished doctoral dissertation (The University of Chicago, Chicago).
1967 "A System of Implicit Quantification", *The Journal of Symbolic Logic* 32, 484-509.
1968 "Peirce's Graphs–the Continuity Interpretation", *Transactions of the Charles S. Peirce Society* 4, 144-154.

SUBJECT INDEX

Abstractions 64ff, 75ff, 81, 85, 90, 105, 114
Actuality 89-91, 102, 105
Aggregation 111
Algebra 13, 15, 17f, 21f, 111, 118ff, 125-126, 130, 131, 132, 134
Alternative proposition 19, 25-26, 28, 29, 125, 129
Analysis 11, 12, 13, 16, 76-77, 95-98, 110ff, 128, 134, 135
Area of a cut 35, 152
Assertion 103
Associated Alpha graph (aag) 145ff
Associated Beta graph (abg) 148
Association 112, 118, 133
Axiom 32, 47, 62-63 (axioms of identity), 84, 119-120, 123, 134, 139, 146
Barbara syllogism 60, 112
Blank 32, 36, 47, 107
Blot 36, 152
Bracket-free notation 118f
Branching line of identity 49
Broken cut 81ff, 152
C1 (Convention 1) 32, 105, 137, 140, 142n.3
C2 32, 105, 137, 140
C3 33, 105, 137
C4 33-35, 137
C5 35-37, 137
C6 47-48, 105, 115, 137
C6+ 105
C7 48-49, 105, 115, 120, 137
C8 49-54, 105, 115, 137
C9 54-55, 137
C10 82, 84, 137
C11 105, 138
Calculus 133, 135
Carnegie Foundation 127n. 15
Categorical propositions 24, 26-27, 29, 47ff, 51-52, 113-114, 122, 133
Categorical syllogism 17, 113ff, 130
Categories 11, 88, 89 and n.3, 93, 98, 100, 103
Character 23-24, 65, 69, 78, 88
Chef d'œuvre 11, 110, 127n. 16, 128
Chemical composition 117-118
Chemical diagrams 17, 21, 22, 25, 117, 130
Class inclusion 130
Close of a cut 35, 152
Closure 146
Commutation 118
Completeness 139-140, 146-151
Composition 111
Concepts 133
composition of, 116-118, 135
Conditional *de inesse* 33, 35, 89, 95ff
Conditional proposition 26, 29, 33, 112
Conditionalization, rule of 120f
Confucian 123
Conjunctive normal form 131
Conjunctive proposition 18, 26, 28, 40, 49, 125, 141
Consequence 33
Consistency 140-144, 145-146
Containment 38 and n.11, 141ff
Continuity 124-125
Contraposition 122
Conversion 122
Cut 27, 31, 34, 35, 88, 125, 152
Deduction 127, 132, 135
Deduction theorem 120, 134
Definition 135
Deiteration 43-44, 58-59, 152
Delta part of EG 135
Destined 93
Diagrams, logic 13, 15, 16ff, 117, 124, 131, 132, 133
Diagrammatic thinking 12, 16ff, 75, 124
Dialogue, all thought is 92, 113, 133, 135
Discourse of mind with itself 92

Disjunctive normal form 130-131
Dotted line 75ff, 91 and n.4
Dotted rim 78 f
Double cut 44, 120n.7, 152
rule of, 44, 59-60, 82
Dyads 115-116
Elementary graph 140, 145
Empty cut 36-37
Enclosure, 35, 152
Endogenous 39n.13
Endoporeutic 39, 51-52, 56, 68, 105, 152
Ens rationis 65, 75, 78
Entire graph 33, 152
Entitative 152
Entitative graphs 17, 25-27, 34, 125, 133
Erasure 41, 56, 112, 152
rule of, 41, 56, 74-75, 112
Essential possibility 86
Evenly enclosed 26, 29, 38, 152
Excluded middle 84
Existence 47, 102, 119-120
Existential 152
Existential graphs
as *chef d'œuvre*, 11, 110, 127-128
date of discovery, 11, 13, 20, 133
purpose of, 11-15, Chapter 2, 31, 92, 110ff, 127-128, 133, 134, 135
Existential import 33, 122, 123
Exogenous 39 n.13
Experimentation 17, 31, 93, 112, 124, 127, 128, 132
Explanation 99
First area 35, 152
First close 35, 152
Formal implication 133
Formal possibility 86
Functional calculus 47, 64, 119, 146ff
General 32
Grand Logic 23, 24, 60
Graph 17, 32, 35, 153
Graph-instance 32, 72, 153
Graphist 32, 91, 92f, 95, 133
Graph-replica 72, 153
Graphs of graphs 64, 71-75, 84
Heavy dash 47ff, 114, 153
Heavy dot, 47, 90, 114, 153
Heraldry 89
Hook 48, 147, 153
Hypostatic abstraction 65f, 80, 108
Hypothetical propositions 113, 133
I Ching 123
Icon 24, 124, 133
Iconicity 17, 68, 100, 112, 123-126
Identity 47, 48, 53-54, 60, 62-63, 125, 132
Indemonstrable implication 47
Individuals 19, 20, 21, 22, 24, 26, 47, 81, 88, 90, 105f, 124-125
Induction 99,127
Infinite class 132
Inherence 24
Inner area 35, 153
Inner close 35, 153
Inner cut 35, 153
Insertion 41-42, 111ff, 153
Instance 153
Intention 93, 102
Interpreter 92f, 114, 133
Iteration 42-43, 57-58, 153
Joint denial 131
Juxtaposition 29, 33, 40, 125, 141ff
Ligature 49ff, 57, 86, 153
Line of identity 20, 21, 22, 26, 29, 47ff, 105, 113, 125-126, 153
Logic 16
Logic Notebook, see Ms 339 in "References to the Peirce Manuscripts."
Logic of propositions 31, 131, 139ff
Logic of quantification 47, 146ff
Logic of relatives 14, 15, 17ff, 23, 25ff, 53, 114, 127, 130, 131, 133, 134
Logical necessity 81ff, 85
Logical possibility 69, 78, 81ff, 88ff, 105
Loop 35
Lowell Lectures of 1903 12, 33, 64, 68, 70, 77ff, 80, 83, 98, 110, 135
ML 139n.1, 146ff
Map 124
Material implication 33-35, 89, 95ff, 108-109, 121, 132, 133
Mathematics, 16, 110, 130
Medad 115-116
Metaphysical necessity 102
Metaphysical possibility 102
Metaphysics 134
Mind as a sign 113
Modality 11, 80ff, 101, 127
Modal logic 27, 64, 135
Modal selective 85-86, 99-101
Modus ponens 45, 139
Natural deduction 120-122, 134
Necessary reasoning 111, 135
Necessity 90, 99, 101, 102
Negation 20, 26, 28, 35-37, 89, 91, 93
Nest of cuts 37f, 153
Node 153
Nominalism 130
n-*th* area 153
Objective possibility 89ff, 97, 106
Observation 16, 112, 124, 130, 132, 135
Obversion 122

Oddly enclosed 26, 29, 38, 153
Omissions (see "Erasure") 111ff
Operations 20
Outer close 35, 153
Outer cut 35, 153
P 139ff
Partial graph 33, 153
Peg 153
Pheme 91
Phemic sheet 91, 92, 93, 102, 105, 113, 124-125, 153
Philonian conditional 89, 95ff
Polyad 115
Place of a cut or graph 35, 38, 54, 153
Positive logical graphs 30
Possibility 95, 101, 135
Posteriority 69f
Potential 68ff
Pragmaticism 11, 89, 98, 127, 135
Pragmatism 95, 126-127
Precisive abstraction 65 n. 3
Predicate 114-115, 133
Predicate calculus 47
Prenex normal form 132, 133
Prescott Book, see Ms 277 in "References to the Peirce Manuscript".
Principia Mathematica 79, 119, 120
Probability 129
Properties, see "Qualities".
Propositional calculus 31, 119ff, 134, 139ff
Propositional function 115
Propositions, classification of 113ff
Province 93, 105
Pseudograph 27, 36-37, 56, 67, 107-108, 153
Qualities 23 (character), 64, 68ff, 81, 105
Quantification, quantifiers 18-19, 24, 26, 29, 51, 55, 83f, 99, 125, 129, 131, 133, 146
R1 (Rule 1) 41, 56, 74-75, 107, 138
R2 41-42, 56-57, 107, 138
R3 42-43, 57-58, 82, 107, 117, 138
R4 43-44, 58-59, 82, 107, 138
R5 44, 59-60, 82, 107, 138
R6 82-83, 138
R7 86, 138
R8 107, 138
R9 107, 138
R10 108-109, 138
Rational necessity 90, 106
Realism 95, 127, 130
Real possibilities 95
Reasoning 16, 92, 111, 112f, 130, 135
Recto 89ff, 117, 153
Reference 88, 89
Relations (see "Logic of relatives") 24, 25, 64, 68ff, 133
Relative terms 22, 154
Relative propositions 29, 113-114
Replica 72, 154
Retroduction 127
Rhema 22, 48, 78, 114-115, 154
Rim 75, 78ff
Rule of necessitation 86
Saw line 78f
Saw rim 71n.6, 78-79
Schema, schemata 139ff
Scribe 31, 154
Scroll 34-35, 154
Second area 35, 154
Second close 35, 154
Selective 50, 55, 69, 75, 77, 85f, 99-101, 115, 154
Self-defence of reasoning 113
Semantical paradoxes 130
Sep 35 n.4, 154
Sequence 69f
Sheet of assertion 31, 80f, 88ff, 105, 154
Sheet of Destination 92
Sheet of Interrogation 91
Signs, theory of 11, 113, 127
Skolem normal form 132n.2
Spot 21, 24, 28, 47ff, 65, 115, 117, 126, 145, 147, 154
State of information 84-86
State of things 99
Stecheology 21
Strict implication 89, 108-109
Subject 114-115, 133
Subjective possibility 90, 95
Substantive possibility 70, 86
Symbolic logic 15, 129, 134
Synechism 127
Tautology 140-141
Teridentity 74, 116
Thought-process 113
Tinctures 28, 89ff, 125, 135
need for, 95, 100f, 104
Token 32
Total graph 33, 154
Triads, irreducibility of 115-116
Truth-functional conditional 33, 39
Truth-functional propositions 113-114
Truth-table analysis 40ff, 131, 134
Type 32
Universal 32
Universe of discourse 31ff, 47ff, 65, 67, 80f, 88ff, 100, 102, 124f, 154

Universe of possibilities 81, 89
Valency 22, 25, 117-118
Value of an area 40f
Verso 34, 88-91, 117, 154
Wavy line 78f
Wavy rim 78f
Well-formed formula (wff) 115, 139, 146

NAME INDEX

Aquinas, Thomas 132n.3
Aristotle 17, 49, 130, 132n.3
Baldwin, James Mark 134
Berry, George D. W. 12, 13, 18n.5, 129, 130n.1, 131, 134
Bochenski, I. M. 12, 130, 131, 132, 133, 134
Boole, George 11, 129, 130
Burks, Arthur 129
Cantor, G. 110
Carus, Paul 98n.9, 132
Church, Alonzo 12, 62, 120, 126, 129, 131, 132, 133, 134, 139, 144
Clifford, William K. 17, 22, 130, 132
Cope, Jackson I. 17n.3
Cresswell, M. J. 86n.16
Crombie, E. James 8, 43n.15
Dedekind, Richard 132
De Morgan, Augustus 17
Euclid 77, 116
Euler, Leonhard 18
Feibleman, James K. 12
Fisch, Max H. 7, 8, 11n. 6, 15, 17n.3, 98n.9, 110, 124n.13, 135
Frege, Gottlob 131
Gallie, W. B. 13
Gardner, Martin 14, 18n.4
Gauss, C. F. 17, 130
Gentzen, Gerhard 120, 134
Goudge, Thomas A. 13
Gray, Chistopher P. 8, 57n.3
Hailperin, Theodore 119n.6
Hintikka, Jaako 119n.6
Hughes, G. E. 86n.16
James, William 11, 14, 79, 134, 135
Jaśkowski, Stanisław 120, 134
Jourdain, Philip E. B. 11n.6, 110, 128
Kant, Immanuel 134
Kempe, Alfred Bray 20-22, 24-25, 60, 72, 132, 133
Kneale, Martha 12, 120, 129, 130, 131, 132, 133
Kneale, William 12, 120, 129, 130, 131, 132, 133
Ladd-Franklin, Christine 119, 134
Leblanc, Hughes 119n.6
Leibniz, G. W. v. 132
Leonard, Henry 119n.6
Lewis, C. I. 129, 130, 131, 132n.2
Łukasiewicz, Jan 118, 120n.9, 135
Mitchell, O. H. 18, 19, 20, 26, 111, 131
Murphey, Murray G. 12, 16n.1, 17, 127n.14, 130
Paulsen, Friedrich 134
Peirce, Benjamin 8, 16, 130
Plato 105
Post, Emil 119, 135
Prior, A. N. 118
Quine, Willard Van Orman 12, 13, 62, 115n.3, 120, 125, 132n.3, 146
Risteen, Allan Douglas 99n.13, 135
Roberts, Don D. 15, 115n.3, 141n.2
Robin, Richard S. 7
Roosevelt, Theodore 101
Royce, Josiah 103
Russell, Bertrand 79n.11, 119, 120, 129, 132
Schafter, Thomas Lee 38n.11
Schröder, Ernst 133
Scotus, D. 89
Sheffer, H. M. 131
Stephenson, David 43n.15
Sylvester, James J. 17, 130
Turquette, A. R. 15, 135
Tursman, R.A. 163
Venn, John 18
Welby, Lady Victoria 99n.12, 117n.4
Wells, Rulon 15
Whitehead, A. N. 79n.11, 118, 119, 120, 129
Woods, Frederick Adams 11n.3, 99n.13, 103, 135
Zeman, J. Jay 15, 86n.16, 115n.3, 127n.16, 147n.5

REFERENCES TO THE COLLECTED PAPERS OF PEIRCE

Note: When a passage is cited in footnotes only on any page, the numbers of the footnotes are given; otherwise, the page number alone is given.

1.23 90n.3
1.28-34 130
  1.34 17
1.191 21n.9
1.346 116
1.549 65n.3

2.27 113
2.121 134
2.182 113
2.184 113
2.279 124
2.323 82
2.334 133
2.347 85
2.382 82
CP 2, p. 2 127n. 14

3.1-19 129
3.20-44 129
3.45-149 17, 130
  3.45 17
  3.92 124
  3.99 17
3.199 111
3.204ff 130
3.252-288 131
3.328-358 131
  3.329 18n.5
  3.351 18n.5
3.359-403 131
  3.363 17, 112, 124n.12, 132
  3.380 120 n. 10
  3.384-385 132
  3.396 132
  3.398 132
3.404-424 132
  3.418 22, 124n.12, 132
  3.419 132
  3.420 22
  3.421 22
3.425-455 133
  3.426-429 133
  3.442 86
  3.442-443 95n.6
  3.443 96n.8, 133
  3.445 133
3.456-552 133
  3.457 25n.10
  3.467 114
  3.468 17, 21, 24, 25n.10
  3.469 25, 117
  3.469-470 17
  3.470 117
  3.479n.1 24
  3.505 132n.2
3.556 124n.12
3.572 88
3.601 21n.9
3.619 134
3.620 15
3.641 124n.12

4.6 112, 113
4.9 21n.9
4.12-20 131
4.67 86
4.80-84 133
4.227-323 134
  4.229 16, 130

4.233	124, 135
4.235	65n.4
4.239	135
4.260-262	134
4.264	134
4.280	111
4.346	65n.3
4.372	15, 134
4.374	111, 118
4.387	51
4.394-417	27n.14
4.395	36n.9
4.395n.1	72, 154
4.396	137, 142n.3
4.397	32, 137
4.398	33
4.399	35, 152
4.401	137
4.402	36n.8
4.404	137
4.405-407	50n.1
4.405	137
4.406	137
4.407	53, 137
4.409	69n.5
4.410	82, 137, 152
4.411	78
4.414	35, 50n.1, 152
4.415	37n.9, 120n.7
4.416	50n.1
4.417	47, 137
4.418-509	64n.1
4.431	92, 142n.3
4.433	137
4.434	125
4.435	35n.4, 137
4.437	35, 137, 152
4.438	115, 154
4.441	49, 154
4.444	137
4.446	137
4.448	125
4.449	50n.1
4.450	137
4.452	36, 53n.2
4.455	36nn.8, 9
4.458	50n.1, 51
4.459	50n.1
4.460-461	55
4.460	55, 154
4.462	55
4.463	65nn.3, 4, 66
4.464	132n.3
4.467	36, 137
4.469	53
4.470	68, 76n.9
4.471	75, 77
4.473	55
4.474	49, 50n.1
4.476ff	40
4.485-498	120nn.8, 10
4.492	138
4.501	54
4.505	57n.3, 138
4.506	138
4.507	76
4.508	138
4.512	68, 81, 124, 125, 142n.3
4.514	81, 125
4,515	84, 137
4.515ff	81
4.516	84, 138
4.517	85
4.518-522	85n.14
4.518	85, 138, 154
4.519	83
4.521	86
4.522	85n.14
4.524	69n.5
4.526	69n.5
4.528-529	71
4.528	71n.6, 73n.7
4.529n.*	71
4.530-572	27n.14, 92, 135
4.530	124
4.537	32n.1
4.546	95n.5, 96, 97
4.547	95
4.549	65n.3, 95n.5
4.553ff	153
4.553n.1	90
4.554	102
4.556	35, 152, 154
4.559	137
4.560	49
4.561	50n.1, 55, 116, 137
4.561n.1	25n.11, 55, 124, 125
4.562	104
4.564-571	91n.4

4.564 36nn.6, 8
4.567 36n.8, 47, 87n.2, 120n.7, 137, 138, 142n.3
4.569 87n.2, 94, 95, 103, 135
4.571 60, 127
4.572 117
4.573-584 135
4.575 89
4.576 88
4,578 88
4.579 50n.1, 95
4.580 95n.5, 96
4.581 57n.3, 88, 154
4.582 113
4.583 117
4.615 17
4.617 14, 36n.8, 99, 103
4.620ff 87n.1
4.621 153
4.622 110
4.653 98

5.147 60
5.313 113
5.318-357 130
5.365 40
5.415 127
5.453ff 127
5.505 21n.9
5.554 98n.9

6.174 21n.9
6.588-618 132
6.590ff 101n.15
6.595 132

7.393 112, 133
7.428 112
7.429 112
7.430 112
7.467 133

CP 8, pp. 251-321 129
CP 8, p. 297 103
CP 8, p. 298 87n.2

REFERENCES TO THE PEIRCE MANUSCRIPTS

Note: When a manuscript is cited in footnotes only on any page, the numbers of the footnotes are given; otherwise, the page number alone is given.

Ms 78 16, 130
Ms 277 (*The Prescott Book*) 53, 101
Ms 280 11n.5, 14, 27nn.12, 13, 32, 104, 114
Ms 283 98n.9
Ms 292 89, 95, 99n.11
Ms 295 90, 91, 93, 94, 98, 102
Ms 296 99n.10, 117n.4
Ms 300 55, 93, 99n.11, 100, 101, 116, 127
Ms 302 127n.14
Ms 330 127
Ms 339 (The *Logic Notebook*) 7, 27n.14, 30, 32, 64, 65, 87n.1, 88, 89, 95, 99, 101, 104, 120nn.7, 8, 121, 124n.12, 134, 135
Ms 410 24, 133
Ms 436 127, 133
Ms 437 15n.11
Ms 439 115
Ms 447-478 64
Ms 450 14, 31, 32n.1, 33, 34, 35, 36n.9, 47, 49, 55, 64n.1, 92, 154
Ms 454 51, 120
Ms 455 31, 33, 34, 47, 48, 49, 55, 111, 120
Ms 458 65
Ms 459 67, 69, 70, 105
Ms 460 65n.4, 81
Ms 462 47, 65n.4, 66, 67
Ms 464(s) 78
Ms 467 66, 69, 71, 73n.7, 81, 86
Ms 468 71, 72, 83
Ms 469 67, 68
Ms 478 69, 70, 78, 82, 87, 98, 120n.8, 147n.6
Ms 481 48, 51, 53, 114
Ms 484 29, 50, 92, 120n.8, 125
Ms 485 28, 30
Ms 488 30

Ms 490 32n.1, 38n.11, 57n.3, 88, 89, 95n.5
Ms 492 55, 64n.1, 75, 76n.9, 77
Ms 493 49
Ms 494 55
Ms 498 11n.3, 20, 27n.12, 32n.1, 110, 116, 133
Ms 499 111, 116
Ms 499(s) 100
Ms 500 11n.3, 14, 27n.12, 60, 133
Ms 503 50
Ms 504 53n.2
Ms 511 72
Ms 513 27n.12, 29, 30
Ms 514 99, 117
Ms 524 124
Ms 620 126, 135
Ms 631 11n.1
Ms 650 27n.14, 36nn.6, 8, 37n.10, 39n.13, 41, 42n.14 43, 92, 99n.12, 124
Ms 669 101n.16
Ms 670 101n.16, 102, 105, 135
Ms 674 102
Ms 693 34n.3, 37n.10, 75n.8, 112
Ms 708 25
Ms 714 22, 132
Ms 715 20
Ms 905 111
Ms 1147 113
Ms 1167 152
Ms 1170 20
Ms 1454 134
Ms 1589 11n.3

Ms L 75 127n.15
Ms L 77 98n.9
Ms L 224 11n.7, 30, 79, 87n.2, 134, 135
Ms L 230a 110
Ms L 231 32n.1, 50n.1, 117, 124
Ms L 237 119, 134
Ms L 294 18, 131
Ms L 376 99n.13, 135
Ms L 463 87n.2, 99n.12, 117nn.4, 5
Ms L 477 11n.3, 27n.12, 29, 36n.8, 39n.13, 87n.2, 99n.13, 133, 135

Ms S 28 68
Ms S 30 36n.8

Ms AM 806* 27n.12, 30, 133
Ms AM 806.5* 60

Bei Fragen zur Produktsicherheit wenden Sie sich bitte an:
If you have any questions regarding product safety,
please contact:

Walter de Gruyter GmbH
Genthiner Straße 13
10785 Berlin
productsafety@degruyterbrill.com